T5-CVS-056

Technology & Economics

Papers commemorating
Ralph Landau's service to the
National Academy of Engineering

NATIONAL ACADEMY PRESS
Washington, D.C.
1991

NATIONAL ACADEMY PRESS • 2101 Constitution Avenue, NW • Washington, DC 20418

The National Academy of Engineering was established in 1964, under the charter of the National Academy of Sciences, as a parallel organization of outstanding engineers. It is autonomous in its administration and in the selection of its members, sharing with the National Academy of Sciences the responsibility for advising the federal government. The National Academy of Engineering also sponsors engineering programs aimed at meeting national needs, encourages education and research, and recognizes the superior achievements of engineers. Dr. Robert M. White is president of the National Academy of Engineering.

This volume consists of papers and speakers' remarks during a commemorative symposium in honor of outgoing NAE Vice President Ralph Landau. The symposium, entitled "Technology and Economics," was held 4-5 April 1990. The interpretations and conclusions expressed in the symposium papers are those of the authors and are not presented as the views of the council, officers, or staff of the National Academy of Engineering.

Library of Congress Catalog Card No. 90-63562
International Standard Book Number 0-309-04397-2

Additional copies of this publication are available from:

National Academy Press
2101 Constitution Avenue, N.W.
Washington, D.C. 20418

S-258

Printed in the United States of America

Contents

* * * * *

The following papers were presented at a Symposium on Technology and Economics in honor of Dr. Ralph Landau for his contributions toward increasing the understanding of interactions of technology and economics.

APPENDIX

Foreword

The papers published in this volume commemorate Ralph Landau's many years of service to the National Academy of Engineering. Over almost two decades, Ralph Landau served the NAE as member of the council, officer of the Academy, and in a variety of capacities as a vigorous intellectual contributor to the Academy's program. Ralph Landau is an innovator who combines the characteristics of a personable and valued colleague with those of a hard-driving leader. The NAE is a different organization for his having volunteered his time and efforts on its behalf. He has taken the lead on issues that range from membership policy, through the NAE's program on technology and economics, to the NAE's relationship with the National Academy of Sciences and the National Research Council. In no small way he was also responsible for the success of the Academy's 25th Anniversary Fund Drive, of which he was chairman.

It has been a personal pleasure working with Ralph Landau and I am pleased that the institution can honor him with this volume. Six of the papers in the volume were presented at a symposium held in Ralph's honor on April 5, 1990, in Washington, D.C. The seventh paper, a contribution by Ralph Landau himself, is part of his continuing effort to build bridges between economists and engineers, to deepen our national understanding of the interactions of technology and economics.

I would like to thank NAE staff members Melvin Gipson, Maribeth Keitz, H. Dale Langford, and Bruce Guile. Mr. Gipson took the lead organizing the symposium in Dr. Landau's honor and worked with Mr. Langford, the NAE's editor, in bringing this book to fruition. Ms. Keitz provided primary support for both the symposium and the publication process. Dr. Guile, the director of the NAE's Program Office who has worked closely with Dr. Landau for the last six years, provided direction and oversight for the project.

Robert M. White
President, National Academy of Engineering

Ralph Landau: Engineer, Entrepreneur, Scholar

Ralph Landau is the former chairman of the Halcon SD Group, Inc. Born in Philadelphia, he received his primary and secondary education there, graduating from the University of Pennsylvania in 1937 with a bachelor of science degree in chemical engineering. Four years later he earned a doctor of science degree in that field at the Massachusetts Institute of Technology.

From 1941 to 1945 Dr. Landau worked as a process development engineer and head of the chemical department of Kellex Corporation, where he engaged in work on the Manhattan Project at Oak Ridge, Tennessee. After the war, he and a partner cofounded Scientific Design Co., Inc., and in the early 1950s developed an original process for the manufacture of terephthalic acid, the key ingredient of polyester fiber. This technology was later sold to Standard Oil Co. (Indiana) and formed the basis for the establishment of the AMOCO Chemicals Co., still the world's largest manufacturer of terephthalic acid.

In time, Scientific Design and its successors (Halcon International, Inc., and Halcon SD Group) became a leading source of modern petrochemical technology in more than 30 countries and owner of some of the most important technology in the chemical industry. In addition to its commercial processing developments, Halcon has designed or constructed more than 300 plants worldwide and signed license agreements with many other countries. Halcon's research and development activities have produced more than 1,400 patents worldwide.

From 1966 to 1980 Halcon participated equally with ARCO in the formation and operation of a major petrochemical company (Oxirane), based on an original process by Halcon for propylene oxide and coproducts. After achieving world sales of a billion dollars from plants located in Texas, the Netherlands, Spain, and Japan, Halcon sold its half interest to ARCO in 1980, where it now forms the core of ARCO Chemical's expanding business.

With the sale of Halcon to the Texas Eastern Corporation in 1982, Ralph Landau assumed a second career, that of scholar. Through his long-standing interest in education and research, he has served as a trustee or a member (and chairman) of visiting advisory committees at several universities, including Massachusetts Institute of Technology, Princeton University, University of Pennsylvania, and California Institute of Technology. He was a trustee of Cold Spring Harbor Laboratory, a director of Alcoa, and Chairman of the American Section of the Society of the Chemical Industry. He is currently a consulting professor of economics and of chemical engineering at Stanford University and a research fellow in the Kennedy School of Government at Harvard University. In these two posts, as well as in the National Academy of Engineering, he is helping to develop a new academic field aimed at understanding the linkages between technology and economic policy and growth. He has coedited three books from these efforts—and several more are in preparation—and more than 120 papers.

Ralph Landau's awards include the National Medal of Technology, for which he was among the first group of recipients. He is one of only five individuals who have received both the Chemical Industry Medal and the Perkin Medal—two of the highest awards in the chemical industry, reflecting his position as a leading technological entrepreneur of this industry. He is also a recipient of the John Fritz Medal awarded by five engineering societies for scientific or industrial achievement. He is a fellow of the American Academy of Arts and Sciences.

Since his election to the National Academy of Engineering in 1972, Ralph Landau's association with the Academy has been one of constant service and leadership. As a councillor from 1973 to 1979 and vice president since 1981, he holds the record for length of service on the Council. From 1984 to 1989 he chaired the Academy's 25th Anniversary Fund Drive. This effort, which yielded more than $46 million (including the establishment of the Arnold and Mabel Beckman Center for the National Academies of Sciences and Engineering), will enable programs that will be a lasting mark of Ralph Landau's contribution to the Academy, to his profession, and to the nation.

How Competitiveness Can Be Achieved: Fostering Economic Growth and Productivity

Ralph Landau

We hear much about the lack of competitiveness of the United States, but seldom is this concept defined, except in terms of international trade balances and market share.[1] It is obvious that this country could improve its trade balance if we reduced the wages and living standards of the American working population to those in Mexico, China, or Brazil, but this would not make America more competitive. What we should mean by competitiveness, and thus the principal goal of our economic policy, is the ability to sustain, in a global economy, a socially acceptable *rate of growth* in the real standard of living of the population with a politically acceptable fair distribution, while efficiently providing employment for those who can and wish to work, and doing so without reducing the growth potential in the standard of living of future generations. This last condition constrains borrowing from abroad, or incurring excessive future tax or spending obligations, to pay for the present generation's higher living standard. As discussed below, such criteria for competitiveness have historically been best realized in industrialized countries by a healthy annual increase in labor force productivity in which the United States has been the leader for most of the past century, and still is in absolute level.

If the U.S. economy could be isolated so that international trade balances were not significant and domestic capital needs were met by domestic savings, these growth criteria for the economy would still apply, but policy could be adjusted more easily to reflect domestic political preferences, such as in targeting interest and inflation rates. Now, with trade in goods and services constituting almost 20 percent of gross national product (GNP), that is, the sum of exports and imports, and the country importing approximately $110 billion of capital in 1989,[2] the previous freedom to set policy is no longer possible. The country must be able to pay for its essential imports (of goods, services, and capital) by exports, and thus international competitiveness and growth in domestic living standards cannot be separated from each other. It is important to examine briefly the changes in the international economy since the Second World War, to understand both this growing economic interdependence among nations and the resulting changes in economists' views of appropriate policy options.

During the first 20-25 years after the war, the United States enjoyed an essentially unlimited economic horizon. Propelled by the head start this situation permitted, real U.S. gross domestic product—GDP—(which differs slightly from GNP by omitting net factor incomes from abroad), tripled since 1950 and income per capita almost doubled; meanwhile real GDP of the world, aided by the United States to recover from the war, quadrupled. The United States relied on domestic savings to meet its domestic capital needs, exported capital to the recovering countries, and used macroeconomic policy to adjust demand to cyclical changes. Supply could—and did—take care of itself through the vigorous activities of the private sector.

World trade in this period grew sevenfold and enhanced this remarkable economic growth. Indeed, systematic empirical research indicates that a closed economy is ultimately a low growth economy (Grossman and Helpman, 1990). There are compensating advantages to greater participation in the world economy, such as the opportunity for nations to specialize in areas particularly advantageous to them, even though other nations have caught up and become strong competitors. Trade permits achievement of economies of scale in strong industries, and raises the level of consumer welfare by providing a greater diversity of goods and services of higher quality. Trade provides greater opportunities for exploiting research successes made in one country in other countries, first by trade and then by local manufacture. These advantages can likely become even larger as the rest of the world becomes more prosperous and provides additional markets for our goods, services, and investments.

Nevertheless, it is clear that the arena of U.S. firms and entrepreneurs has irrevocably changed. International capital and technology flows have become global and in many cases virtually instantaneous. Therefore, domestic freedom to control national destinies, formerly taken for granted, is increasingly constrained by the disciplines of the international capital markets, as well as by the trade in goods and services. On the other hand, fiscal and monetary policies, as well as those dealing with trade, legal, tax, financial, and other matters, vary widely among countries.

At the same time, the world continues to develop extraordinary new technologies that promise to substantially raise global living standards. The age of the computer has just started, but it has already penetrated widely (Figure 1). Telecommunications via satellite and fiber optics are binding the world together at an ever-increasing rate. Robotics provide the means to eliminate hazardous and boringly repetitious tasks. The biotechnology revolution has hardly begun, but already its potential to affect human health and improve productivity in farm and factory is immense. Superconductivity is certain to play a major role in the twenty-first century; new materials are penetrating realms as diverse as medicine and aerospace; new catalysts and pharmaceuticals are improving the efficiency of industry and the human body. Many of these developments are American. To be a scientist or technologist today is to be at the frontier of human explorations

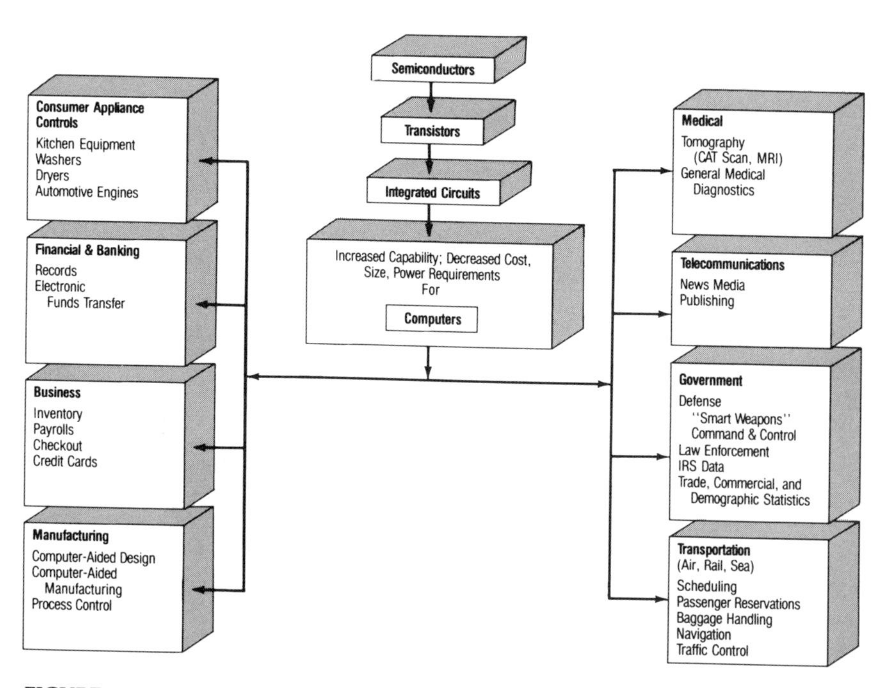

FIGURE 1 The impact of technology on economic development: new processes, products, and services. SOURCE: *The Technological Dimensions of International Competitiveness.* Prepared by the Committee on Technology Issues that Impact International Competiveness, National Academy of Engineering, Washington, D.C., 1988, p. 14.

and aspirations, but we must be cognizant of the economic and social limitations on such exciting prospects. What in this should or does make us worry?

From 1870 to 1984, the country's average real growth rate in GDP, was about 3.4 percent per year; from 1948 until recently it exceeded this level. This growth was accomplished mainly by a growth rate of about 2 percent per year in real income per person in the United States and the rest by average growth rate in population. Standards of living nearly doubled between generations. The United States surpassed the United Kingdom, the one-time leading industrial power, whose *per capita* real income grew at only 1 percent per year. Today, the United Kingdom is not even the leading member of the Common Market. On the other hand, starting with the Meiji restoration of 1868, Japan has recently exceeded even the high American growth rate. With an annual real GDP growth rate of more than 6.9 percent from 1952 to 1987, it has become the second largest economy in the world.

Such dramatic reversals underscore the power of compounding over long periods of time. Differences of a few tenths of a percentage point, which may not appear very significant in the short term, are an enormous economic and social achievement when viewed in the long run. For example, an increase of only 2.5 percent in the annual growth rate (which means raising the growth rate of GDP by less than 0.1 percentage point per year) will double the standard of living per capita in less than 30 years (a generation) with a constant population. Thus, it is of concern that since 1979 the U.S. real annual GDP growth rate has averaged only about 2.75 percent, with substantial year-to-year fluctuations, and with an almost static per capita real income, despite a more than seven-year economic recovery. Will the United States follow the fate of the United Kingdom, while Japan and the Far East, or post-1992 Europe (aided by the appearance of new markets in Eastern Europe) eventually outdistance it? Or can it maintain a prominent stance of economic, and strategic leadership, which its unique position of being *both* an economic and a military superpower demands of it? The answer to this question is not at all clear, and this alone is a reason for worrying.

Both economic evidence and historical experience suggest that sustained economic growth does not come from only doing more of what we already do, although in a global economy we must capitalize on existing technologies more fully and more rapidly than ever. For 100 years, our economy grew because we made the capital investments necessary to exploit great discoveries such as machine tools, the electric motor, petroleum exploration and refining, and semiconductors, to name but a few. Our poorer recent growth performance cannot be attributed to a dearth of new investment opportunities now. To achieve more rapid economic growth, the promise of the new technologies must also be realized, but it cannot be accomplished without taking into account the historical realities under which new technology is applied.

THE ROLE OF TECHNOLOGICAL CHANGE IN GROWTH

The United States could have achieved its growth in per capita real income (1) by using more resources, or (2) by getting more output from each unit of resources (increasing the productivity). How much of the long-term rise in per capita incomes is attributable to each? The surprising answers emerging in the 1950s indicated that long-term economic growth (since the Civil War) had not come from simply using more and more resources, that is, capital and labor, but rather, overwhelmingly (85 percent) from using resources *more efficiently*. Many attached the label "technical change" to that entire *residual* portion of the growth in output which cannot be attributed to the measured, weighted growth in inputs and thus equated it to the growth in productivity. Certainly, however, many social, educational, and organizational factors, as well as economies of scale and resource allocation, also affect productivity. Stanford's Moses Abramovitz (1956), who published some early studies of this nature, called it "a measure of our ignorance."

Out of this work came the detailed growth accounting studies of the 1960s and 1970s, based on the neoclassical growth theory of Robert Solow (1957) at the Massachusetts Institute of Technology. This theory holds that in a perfectly competitive economy, in the long-run steady state, the rate of growth is independent of the saving (or equivalently the investment) rate; in other words, growth is independent of the proportion of output that is reinvested. These studies, led by Edward Denison of the Brookings Institution, Zvi Griliches and Dale Jorgenson of Harvard University, and John Kendrick of George Washington University, sought to reduce the residual by identifying some of its components and measuring the inputs more accurately.

In all these studies, the strangest aspect was that the actual sizable growth rates in the industrial countries constituted a remarkable economic phenomenon: a tribute to the dynamic performance of capitalist economies, especially significant in view of the collapse of the socialist economies in the 1980s and the reevaluation of Soviet growth rates to an essentially stagnant or declining level. And yet, because the technology, the residual, was assumed in this theory as *exogenous*, not a product of traditional economic activity, it appeared that a large part of this remarkable accomplishment was unknowable, generated somewhere outside of the economy!

Economists responded to this challenge by studying the American economy from various perspectives. Some of the group mentioned above tried to relate technological change to economic forces, and thus sought in effect by various approximations to endogenize or integrate the measured phenomena into the rest of the economy. The residual was thus a summary, at the aggregate or *macroeconomic* level, of forces occurring at the micro level of firms and individuals, and was therefore really a part of the economy. However, the unexplained part remained disturbingly large and variable, and there were many assumptions and intuitive elements involved.

Another version of this approach addressed the measurement issues, on the assumption that if the economic variables such as scale economies and the quality and quantity of inputs were properly measured, the residual could be greatly shrunk. Obviously, as better data and methodologies became available in more recent years, this work, described in the recent book by Jorgenson and Landau, *Technology and Capital Formation* (1989), did shrink the residual, but it did not go away. Some of this may have been due to still unrecorded measurements such as the acquisition of human capital, to various social and political factors, and also, as we have shown in our detailed study of the chemical process industries (Landau, 1989a, 1990b; Rosenberg and Landau, 1989), to less-than-capacity utilization at times, which has had a very negative influence on productivity. Certainly no methodology of this kind is free of assumptions either, although there are fewer of them.

Nevertheless, by either approach, a significant residual remains and has difficult-to-explain large fluctuations at intervals. Furthermore, the extrapolation by these methods from the infinite variety of microeconomic activities of firms to the macro economy over time was either a rather bold leap of faith, or else the models developed were too simplistic to reveal the functioning of the "black box" of technical change at the firm level, and so left obscure just what could or should be done to increase growth rates, which, after all, is the point.

In fact, a major assumption of present-day neoclassical macroeconomists about the microeconomic world, is the textbook assertion that business firms are homogeneous maximizing agents, whose history, internal structure, and characteristics are not examined, or at least are not central to the analysis. Such a static view holds that eliminating inefficiencies and gaining economies of scale are the keys to success.

This treatment is a requirement for their growth analysis at the macroeconomic level but, in so doing, they virtually throw away the essential elements of the problem of technology commercialization. They also disregard how firms can be managed for greater competitiveness in the international marketplace—a far more powerful growth mechanism than the static efficiency model, because it continually introduces new products, processes, and services to disrupt any supposed steady state. If the economics textbooks are right, why is the business and general literature so full of accounts and advice of how different firms and industries are succeeding or failing in the international and domestic markets? And, with all their imperfections, the capital markets recognize their varying results. This is the puzzle that conventional growth theory cannot solve.

Within the past few years, new international phenomena have begun to draw the attention of economists as a means of widening their understanding of the growth process. In the past two decades, some fascinating divergences in growth rates have occurred outside the socialist bloc, such as the swift rise of the Asian "dragons," the economic decline of some South American and African countries, and, above all, the extraordinary recovery of Japan from wartime devastation. In addition, there was an almost universal slowdown in growth in the 1970s, with some recovery in the 1980s. Meanwhile, advances in economic theory were taking place. Kenneth Arrow of Stanford as early as 1962 had already pointed the way toward a better understanding of this issue. If the predicament of exogenous technical change was to be escaped, and the possibility of sustained and fluctuating growth per capita (as actually occurred) was to be retained, there has to be some form of nonconvexity in the production process, aided by endogenous technical change. From such international observations, development economics and growth theory seemed to begin to merge.

This line of work has recently been led by Robert Lucas (1988) and Paul Romer (1986; 1987a,b; 1989a,b; 1990) of the University of Chicago, but it has received varying amounts of support in papers and statements delivered at Robert Solow's sixty-fifth birthday symposium in April 1989 at the Massachusetts Institute of Technology. This support came from Joseph Stiglitz and Robert Hall of Stanford University, Frank Hahn of Cambridge University, and Avinash Dixit of Princeton University. In addition, Richard Nelson (1981, 1982) of Columbia University has produced a new evolutionary theory of growth that has many similarities to the new work of Lucas and Romer. Gene Grossman and Elhanan Helpman (1990) of Tel Aviv are also supportive of these new directions in growth and trade theory. This new work in growth theory reignited interest in increasing returns to scale as one of the forces driving growth, especially for less developed economies, and introduced complex general equilibrium models into growth research. But economies of scale are also important for industrial countries, particularly for industries in which American firms are strong, such as aircraft, chemicals, machinery, and motor vehicles (Lipsey, 1990). The residual disappears but is replaced by the postulate of externalities, or spillovers—that is, the influence of investments of all kinds on one another. These models also include imperfect competition as the only form that can allow a role for patents and privately financed R&D, as actually occurs. This work brings back into growth theory some of the key concepts first disclosed by Evsey Domar of the Massachusetts Institute of Technology and Roy Harrod of Oxford University (Eatwell et al., 1987) even before Solow's publications, although of course in a far more sophisticated manner.

Our practical observations of the economy would support such a concept. In our studies of the petroleum and chemical industries, we describe how the invention of the assembly line by Henry Ford led to the development of modern petroleum refining aided by the rise of the chemical engineering discipline, which in turn led to the great expansion of the chemical and petrochemical industries, first in the United States, and then abroad. The penetration of the computer has had comparable if not even greater effects. Jeffrey Bernstein of Carleton University and M. Ishaq Nadiri of New York University measured such spillover effects for the high-tech industries and found them to be substantial in almost every case for R&D capital (Bernstein and Nadiri, 1988). They also measured rates of return on both physical and R&D capital, and showed that the latter are higher. However, it is very hard to incorporate the detailed micro view into these models, and much remains to be done.

From the recent work in growth theory, we therefore perceive two important modifications in Solow's neoclassical growth theory, which affect both economic research and its policy implications: (1) it applies to long- run steady-state equilibrium of the economy and not necessarily for the more immediate challenges in periods of less than perhaps 25 or 50 years because the economy in such periods is in a dynamic transition disequilibrium stage; and (2) technology in a mature economy like the United States is largely endogenous.

Other deficiencies of the neoclassical theory, in our view, lie in the omission of public, environmental, and R&D capital stocks, the growing openness of the economy and trade, premature technological obsolescence from external shocks, the different vintages of capital stock, which are not perfectly substitutable for one another, and the not necessarily constant returns to scale in production. Markets are not always perfectly competitive as the neoclassical theory postulates; rather the competition is more often Schumpeterian (innovative, entrepreneurial), and this is a much more powerful force for growth than standard classical price competition. Firms have found, particularly in an era of international competition, that price wars are unattractive, and seek to focus, where possible, on those forms of competition for which there are greater potential profits, that is, the development of new products and processes. There are, of course, many commodity markets that are price-competitive. However, particularly in those manufacturing industries shown in Table 1, managerial energies seek to differentiate themselves by product distinctions, better and lower-cost technologies and operating procedures for their production, and more successful financing strategies. These are research-intensive industries that collectively perform 95 percent of all the industrial R&D in the United States, industries in which there is continual introduction of new products and rapid technological change. Robert Hall of Stanford University, in a timely National Bureau of Economic Research reprint (1092) has studied pricing versus marginal cost in a number of American industries. He shows that American firms often sell at prices well

TABLE 1 The Major R&D Investment Industries, 1989 Estimates (More than $1 Billion)

	R&D Expenditures (in billion $)		
Industry	Total	Privately Financed	Percentage Privately Financed
1. Aerospace	19.16	3.45	18
2. Electrical Machinery & Communications	18.55	10.57	57
3. Machinery	12.13	10.43	86
4. Chemicals	11.52	11.17	97
5. Autos, Trucks Transportation	11.41	9.47	83
6. Professional & Scientific Instruments	6.52	5.54	85
7. Petroleum Products	2.09	2.07	99
8. Rubber Products	1.24	0.93	75
9. Food & Beverages	1.17	1.17	100
TOTAL	83.79	54.80	

NOTE: Total U.S. R&D estimated at $129.2 billion, of which all industrial R&D is $92.7 billion (67% comes from companies and the rest from government so that the above are the bulk of the investors in R&D).

SOURCE: Battelle Memorial Institute.

above marginal cost, and this fact requires interpretation in terms of theories of oligopoly and product differentiation. He concludes by saying that the evidence against pure competition is reasonably convincing.

Our recent detailed study of the American chemical process industry bears out Hall's conclusion and illustrates the richness of these motivations, and the highly successful resulting growth on a world scale, which have given this industry a consistent postwar positive balance of payments. It is one of two such major manufacturing industries (the other being aerospace). Not all industries have been equally successful, as our research shows, and this exemplifies the problem of dealing with growth at the aggregate or macro level only.

Because of such theoretical limitations, comprehension of changing trends in growth from decade to decade requires comparative empirical studies among nations over shorter periods of time, as a guide to national policies.

GROWTH IN THE UNITED STATES VERSUS JAPAN

First, let us examine the relative performance of the United States and Japan, where the contrasts are the most revealing. Since the mid-1960s, productivity growth in the United States has greatly diminished from previous levels. For the period 1964-1973, the labor productivity growth of the U.S. economy was 1.6 percent per year; but from 1973 to 1978, it fell to -0.2 percent, and in 1979-1986 revived to only 0.6 percent. The Japanese labor productivity growth rates for the same periods were 8.4, 2.9, and 2.8 percent per year, respectively. In much of the later part of this period, the growth of total output in the United States was brought about almost entirely by increases in supply of capital and labor, especially (in the 1970s) the latter, as the baby boom peaked. Although explanations for the collapse in American productivity vary, it seems clear from our recent studies that one of the major reasons is that the comparative performance of the U.S. and Japanese labor productivity growth rates over this period has been heavily influenced by *the much higher (often twice as high) rate of Japanese capital investment* in a number of their industrial sectors, made possible by the very high Japanese savings rates. This is a significant departure from the neoclassical growth model which, as stated above, treats growth as independent of the investment rate. As a result of the low interest rates available in Japan, the discount rate for research and development and other technology-intensive efforts was also low, encouraging long time horizons, as further described below. This helped fuel the rapid adoption by most Japanese industries of the latest available technologies from abroad. Many U.S. industries were not incorporating new technology with the same urgency.

Other reasons uncovered by our work include the two oil shocks of the 1970s (which had a worldwide negative impact on growth rates); the sharp inflation of the 1970s, which gave false signals to managements about market opportunities; the entrance of the baby boomers and other new and less skilled workers; the excess capacity in many industries; and so on. Because this flood of labor market entrants was comparatively cheap, managements favored labor over capital. The ratio of capital to labor in the United States had grown by 3 percent between 1948 and 1973, but then it slowed to less than 2 percent. Growth in Japan's ratio was higher.

The post-1973 decline in growth was not limited to the United States and Japan, but was widespread and variable among many other countries. Now, despite the lower energy costs, most countries have not recovered all the way from the pre-1973 conditions, for a variety of individual reasons, including the time lags needed to adjust to the seismic economic changes of the last two decades, as we discuss later. In studying these many events, we have found that physical capital formation has contributed far more significantly to longer-term economic growth than earlier estimates had suggested. And the residual of technological change, while not wholly explicable by our methodology, constitutes less than 30 percent, rather than the earlier estimates of 85 percent. Of course, like others in the past, we assumed the major inputs to be independent of one another; as we shall see, this assumption needs modification. There are still many measurement issues and methodologies to be resolved, but the direction now seems well supported. The important point of these findings is not their exact magnitude, but that there are several primary identifiable ways to improve growth rates over the medium term of 20 to 30 years: physical capital investment, improvement in labor quality, and R&D and technology.

THE JORGENSON ANALYSIS OF THE SOURCES OF ECONOMIC GROWTH

In the accompanying table (Table 2) we present an analysis of the sources of U.S. economic growth, still employing the neoclassical framework, but improving the methodology for measurement and allocation of inputs. The output of the U.S. economy at the aggregate level is defined in terms of value added for the domestic economy. The growth of output is decomposed into the contributions of capital and labor inputs and growth in productivity. Growth rates for the

TABLE 2 Aggregate Output, Inputs, and Productivity: Rates of Growth, 1947–1985

Variable	1947-1985	1947-1953	1953-1957	1957-1960	1960-1966	1966-1969	1969-1973	1973-1979	1979-1985
Value-added	0.0328	0.0529	0.0214	0.0238	0.0472	0.0360	0.0306	0.0212	0.0222
Capital input	0.0388	0.0554	0.0401	0.0229	0.0367	0.0437	0.0421	0.0392	0.0262
Labor input	0.0181	0.0251	0.0037	0.0124	0.0248	0.0226	0.0128	0.0219	0.0146
Contribution of capital input	0.0145	0.0215	0.0149	0.0083	0.0142	0.0167	0.0149	0.0140	0.0098
Contribution of labor input	0.0112	0.0153	0.0022	0.0077	0.0151	0.0140	0.0082	0.0139	0.0089
Rate of productivity growth	0.0071	0.0160	0.0043	0.0078	0.0179	0.0053	0.0074	-.0067	0.0034
Contribution of capital quality	0.0058	0.0126	0.0069	0.0016	0.0053	0.0058	0.0054	0.0045	0.0022
Contribution of capital stock	0.0088	0.0090	0.0080	0.0067	0.0089	0.0108	0.0095	0.0095	0.0077
Contribution of labor quality	0.0039	0.0060	0.0038	0.0084	0.0041	0.0030	0.0018	0.0024	0.0026
Contribution of hours worked	0.0073	0.0093	-.0016	-.0007	0.0110	0.0110	0.0065	0.0114	0.0063
Rates of sectoral productivity growth	0.0088	0.0142	0.0083	0.0112	0.0190	0.0060	0.0097	-.0012	0.0029
Reallocation of value added	-.0019	0.0007	-.0044	-.0021	-.0021	-.0007	-.0023	-.0053	0.0006
Reallocation of capital input	0.0005	0.0003	0.0013	0.0005	0.0009	0.0001	0.0006	-.0001	0.0009
Reallocation of labor input	-.0003	0.0009	-.0009	-.0019	0.0001	-.0002	-.0005	-.0000	-.0010

SOURCE: Jorgenson and Fraumeni (1990).

period 1947-1985 are given for output and the two inputs in the first column of Table 2. Value added grows at the rate of 3.28 percent per year, while capital grows at 3.88 percent and labor input grows at 1.81 percent.

The contributions of capital and labor inputs to the growth of output are obtained by weighting the growth rates of these inputs by their shares in value added. This produces the familiar allocation of growth to its sources. Capital input is the most important source of U.S. economic growth by a substantial margin, accounting for 44.2 percent of growth during the period. Labor input accounts for 34.1 percent of growth. Productivity growth accounts for only 21.6 percent of U.S. economic growth during the postwar period.

The findings summarized in Table 2 are not limited to the period as a whole. In the first panel of the table we compare the growth of output with the contributions of capital and labor inputs and productivity growth for eight subperiods—1947-1953, 1953-1957, 1957-1960, 1960-1966, 1966-1969, 1969-1973, 1973-1979, and 1979-1985. The end points of the periods identified in the table, except for the last period, are years in which a cyclical peak occurred. The growth rate presented for each subperiod is the average annual growth rate between peaks. The contributions of capital and labor inputs are the predominant sources of U.S. economic growth for the period as a whole and all eight subperiods.

We have found that the contribution of capital input is the most significant source of output growth for the period 1947-1985 as a whole. The contribution of capital input is also the most important source of growth for seven of the eight subperiods, while productivity growth is the most important source for the subperiod 1960-1966. The contribution of capital input exceeds the contribution of productivity growth for seven of the eight subperiods, while the contribution of labor input exceeds productivity growth in the last four of the eight subperiods.

In 1985 the level of output of the U.S. economy stood at more than three times the level of output in 1947. Our overall conclusion is that the driving force behind the expansion of the U.S. economy between 1947 and 1985 has been the growth in labor and capital inputs. Growth in capital input is the most important source of growth in output, growth in labor input is the next most important source, and productivity growth is least important (but far from trivial). This perspective focuses attention on the mobilization of capital and labor resources rather than emphasizing advances in productivity, as is sometimes done by those who primarily favor increased R&D efforts.

The findings we have summarized are consistent with a substantial body of research. For example, these findings coincide with those of L. R. Christensen (Wisconsin) and Jorgenson (1969, 1970, 1973) for the United States for the period 1929-1969. Angus Maddison (1987) (Groningen) gives similar results for six industrialized countries, including the United States, for the period 1913-1984. Assar Lindbeck (1983) (Stockholm) generally concurs. However, these findings contrast sharply with those of Abramovitz, Kendrick, and Solow, who emphasize productivity as the predominant growth source. At this point it is useful to describe the steps required to go from these earlier findings to the results we have summarized.

The first step is to decompose the contributions of capital and labor inputs into the separate contributions of capital and labor quality and the contributions of capital stock and hours worked. Capital stock and hours worked are a natural focus for input measurement, since capital input would be proportional to capital stock if capital inputs were homogeneous, while labor input would be proportional to hours worked if labor inputs were homogeneous. In fact, capital and labor inputs are enormously heterogeneous, so that measurement of these inputs requires detailed data on the components of each input. The growth rate of each input is a weighted average of the growth rates of its components. Weights are given by the shares of the components in the value of the input.

The development of measures of labor input reflecting heterogeneity is one of the many pathbreaking contributions by Denison to the analysis of sources of economic growth. The results presented in Table 2 are based on the work of Jorgenson, Frank M. Gollop (Boston College), and Barbara M. Fraumeni (Northeastern University) (1987). They have disaggregated labor input among 1,600 categories at the aggregate level, cross-classified by age, sex, education, class of employment, and occupation. These data on labor input have incorporated all the annual detail on employment, weeks, hours worked, and labor compensation published by the Bureau of the Census in the decennial "Census of Population" and the "Current Population Survey."

Our measures of capital input involve weighting of components of capital input by rental prices. Assets are cross-classified by age of the asset, class of asset, and legal form of organization. Different ages are weighted in accordance with profiles of relative efficiency constructed by Charles R. Hulten (University of Maryland) and Frank Wykoff (Pomona College) (Hulten and Wykoff, 1981; Hulten et al., 1989; Wykoff, 1989). An average of 3,535 components of capital input are distinguished at the aggregate level. Similarly, the data on capital input have incorporated all the available detail on investment in capital goods by class of asset and on property compensation by legal form of organization from the U.S. national income and product accounts.

The growth rates of capital and labor quality are defined as the differences between growth rates on input measures that take account of heterogeneity and measures that ignore heterogeneity. Increases in capital quality reflect the substitution of more highly productive capital goods for those that are less productive. This substitution process requires investment in tangible assets or nonhuman capital. Similarly, growth in labor quality results from the substitution of more effective for less effective workers. This process of substitution requires massive investments in human capital.

In the Abramovitz-Kendrick-Solow approach, the contributions of growth in capital and labor quality are ignored, since inputs are treated as homogeneous. The omission of growth in labor quality destroys the link between investment in human capital and economic growth, while the omission of growth in capital quality leads to drastic underestimation of the impact of investment in nonhuman capital on economic growth. The results we have presented which involve two different effects, one of measurement and the other of composition or aggregation, reveal that the assumption of homogeneous capital and labor inputs is highly misleading.

We find that growth in the quality of capital stock accounts for two-fifths of the growth of capital input during the period 1947-1985. This quantitative relationship also characterizes the eight subperiods. For the period as a whole we find that the growth of hours worked exceeds the growth of labor quality. However, the growth in hours worked actually falls below the growth in the quality of hours worked for the period 1953-1960. For the period 1960-1985 the contribution of hours worked accounts for almost two-thirds of the contribution of labor input. The relative proportions of growth in hours worked and labor quality are far from uniform.

There is a further complication in understanding the causes of growth, however; quantitative measures of productivity do not fully describe the performance of any economy. Quality of products and services is also of great importance, as the Japanese have notably shown us, but is very difficult to measure (David, 1990). Another measure of productivity growth that is not incorporated into conventional measurements is functionality. The semiconductor industry today sells a million-transistor circuit completely interconnected for the same price that it sold a single transistor some 30 years ago. Thus, the functionality has increased conservatively a millionfold. By next year, a small box containing the new Intel microprocessor (860), costing perhaps $10,000 will have about two-thirds of the performance of a Cray-1 Supercomputer, costing many millions of dollars. But because of such steady cost reductions, and therefore pricing, the value of production in dollar terms understates the quality improvement, and productivity improvement is understated. The same is true in other industries, such as chemicals.

RECENT NEW RESEARCH IN GROWTH ECONOMICS OF INDUSTRIAL COUNTRIES

The current revival of interest in growth economics has been further aided by the award of the 1987 Nobel prize in economics to Robert Solow, who has recently expressed his own reconsideration of the role of capital formation in long-term growth. He has stated that he feared one implication of his theory, that the long-term steady-state rate of growth is independent of the savings rate (or equivalently the investment rate), might have been carried too far with regard to the short and medium term in much of the subsequent economic literature and in government policies—resulting in a downplaying of the importance of capital. "You can't take an old plant and teach it new tricks," he said. Indeed, our experience demonstrates that much of the capital spent by companies to maintain their physical facilities incorporates new technology, so that calculations based only on net capital additions underestimate the driving force of technology in the growth process. Old plants have old technology.

There has been a significant shift in composition of investment from longer-lived to shorter-lived assets, such as computers, which depreciate more rapidly (often in three years). Gross investment data are not affected by such compositional shifts. The substitution of more highly productive capital goods, embodying the new technology, for those that are less productive, improves the quality of capital. The productivity of the economy would thus rise even if net investment were zero. Capital's contribution to growth, taking these quality improvements into account, is accordingly much greater than is generally recognized.

It is therefore incorrect to focus primarily on increasing R&D efforts, important though these are, because the physical capital required to realize the R&D results is usually greater than the cost of the R&D involved, depending on the industry (the proportion of R&D expenditures in manufacturing is rising and now approaches 70 percent of physical capital expenditures; part of the reason for this is the declining rate of growth in physical capital investment.) The U.S. economy is not operating everywhere at the technological frontier, and some industries are, in fact, far behind other nations. Even under neoclassical assumptions, additional investment can produce a longer-term increase in productivity growth if an economy is not at the frontier. What is true for a country is also true for an industry or firm. Once a technological lead is exploited by early market penetration, later entrants, even if possessing better technology, often cannot overcome the first entrant's economies of scale and learning curve improvement. Thus it may take 20 to 30 years of steady investment before existing or potentially important technologies can be fully exploited by American companies, but meanwhile GDP may double, as happened in Japan from 1960 to 1980.

One of the disadvantages of the neoclassical model of technology, capital, and labor is that it focuses attention only on the relative proportions in which these three inputs are used. It does not emphasize the importance of variation in their common rate of growth. There is no question that relative proportions matter. The experience of the centrally planned economies has clearly demonstrated that massive increases in physical capital that are not accompanied by improvements in the technology and the quality of the labor force lead to rapidly diminishing returns, just as the neoclassical model would suggest. But because it treats improvements in the technology and labor quality as being unaffected by public and private decisions, the neoclassical model fails to emphasize that, as we have found, these three inputs are intertwined pieces of the *same process*—a three-legged stool of physical, intangible, and human capital. The latter expression is shorthand for training and education of the work force, but obviously, it does not mean that everyone should be educated to the postgraduate level! Selectivity is an essential element of this leg of the stool, and more is not necessarily better. Intangible capital is not just R&D but also includes design engineering, experimental production, worker training, market development, sustained losses in initial operation and market penetration, legal precautions, and insurance.

In this sense, directly and through its stimulus to and interaction with the other factors of production, technological change has been and is central to U.S. economic growth. In the past, the successful entrepreneurial exploitation of new technologies in the private sector to create new and improved products, processes, and businesses has been a distinctive American characteristic and comparative advantage. In a world of dynamic and ever- changing national comparative advantages, it is important to build on our strengths. We can no longer depend on differences in resource endowments (as Saudi Arabia depends on oil), but must rely on endogenous leads or lags among firms and industries of the industrial nations.

These considerations now explain the results of recent studies by our colleagues at Stanford, who have applied some of the advanced tools available today to the comparative study of growth among a number of the industrialized economies. By so doing, some of the previously mentioned constraints of the Solow model can be relaxed. What were their principal findings?

1. Using modern time series methodologies, Steven Durlauf (1989) has examined business cycles and long-term growth in a number of major industrial countries. He has found little evidence of convergence in these economies. Further, by developing a general equilibrium model of demand and supply complementarities, he has shown that appropriate stabilization policies are not meaningfully distinct from a high growth policy. High investments in one industry can induce high investments in other industries, leading in turn to higher growth rates, which then permit still more investment while increasing present consumption at an acceptable rate. But as *The Economist* (23 September 1989) pointed out in its editorial commentary on these results, they also suggest that the lack of investment not only causes a loss of productive capacity in its own right, but also hurts the value of investments already made. This feedback loop of negative externality well comports with our own industrial experiences, and often leads to excess capacity in one industry, and our measured low productivity therein, as happened in chemicals. Our detailed study of the chemical process industries and other high-tech industries demonstrates that industry level measurements are not only feasible, but are also meaningful and correspond to the actual events. These studies thus lead to the conclusion that stable government policies (fiscal, monetary, trade, and tax) favoring high investment rates may be essential for higher levels of economic well-being in the short as well as the long run.

In fact, this seems to have been the German and Japanese secret of their remarkable economic progress from total ruin at the end of World War II. The well-known German economist Kurt Richebacher, in the May 1990 issue of his newsletter on "Currencies and Credit Markets," summarizes this important point as follows: "Restraint in government spending, wages and consumption has paved the way for rising profits and surging capital spending, those being the drastically improved structural features of Continental Europe. While the Anglo-Saxon countries trumpeted and preached supply-side rhetoric, it was only Continental Europe that put these policies into practice."

Growth paths are distinct for each country, and there are multiple, more or less optimal equilibrium paths, which depend on the ability of any country's system to manage its own affairs. This means that history matters—growth is path dependent. Our study of the rise of the chemical engineering discipline in the United States, in association with the petroleum and chemical industries illustrates this dependence, which led to American primacy in an industry (chemicals) that had for long been dominated by Germany. Other industries have very different historical paths. For example, aerospace is heavily dependent on government for research funding and purchases of military equipment. Convergence of growth paths between countries or industries is not an inevitable process, and in many of the less developed countries, it is rare to find convergence in growth rates with the more industrialized countries. Lucas, Romer, and Stiglitz have emphasized that conventional economic growth theory has paid insufficient attention to explaining adequately such differences in economic growth paths among countries. Some countries can grow better through learning by doing than others, for example, and this is aided by the educational level.

2. Michael Boskin and Lawrence Lau (1990) found from studies of the performance of five industrialized nations that technological change is capital augmenting, and the benefits of technological progress are higher when more capital investment is deployed per worker. What they attribute to technological progress includes what others, as mentioned above, may attribute to improvement in the qualities of the inputs. They estimate that capital and technological change combined contribute about 75 percent of the U.S. growth of output. *Thus, capital formation and technological change are complementary to each other.* This is what we described earlier as the intertwining of the inputs to growth. Their methodology captures the second order or interactive effects not derivable from studies of a single country, which would generally yield first-order or additive effects only. Increased physical capital investment per worker, they show, can raise the rate of productivity gains and enhance competitiveness. Raising the *level* of output (and income) by increased capital investment per worker is a worthwhile goal in itself, but the prospect of raising the *rate* of productivity growth thereby, which the newer theories and research imply, is even more exciting and is a far more powerful force for increasing standards of living.

The United States has invested proportionally less in gross nonresidential physical capital investment than the other major industrial countries for 25 years, even though the level of its gross capital investment has not varied a great deal over the years. Furthermore, with capital augmenting technological change, a steady state may not exist even under neoclassical conditions; the limits to growth are expandable. The first public disclosure of these results by Boskin (whose work at Stanford was finished by Lau) in April 1990 at the National Academy of Engineering symposium "Technology and Economics" (see Boskin and Lau, this volume) was made in a speech also describing the Bush administration's economic priority as the achievement of the highest possible sustainable growth rate, and referred to capital formation and technology as important elements.

We show some of our own related findings in Figure 2, which, taken together with the work of Durlauf, and Boskin and Lau, as well as the research described earlier, can no longer permit any doubt that Solow is right—physical capital investment does have a significantly greater effect on productivity growth in such periods of time as a quarter of a century (or perhaps longer) than the neoclassical long-term growth theory would predict. Boskin and Lau find high augmentation rates for capital in the five countries and a low elasticity of output with respect to augmented capital. These findings are consistent with those of Figure 2 if one converts the horizontal axis from the rate of growth of physical capital to augmented capital per worker. Therefore, the factors affecting the availability and cost of capital in

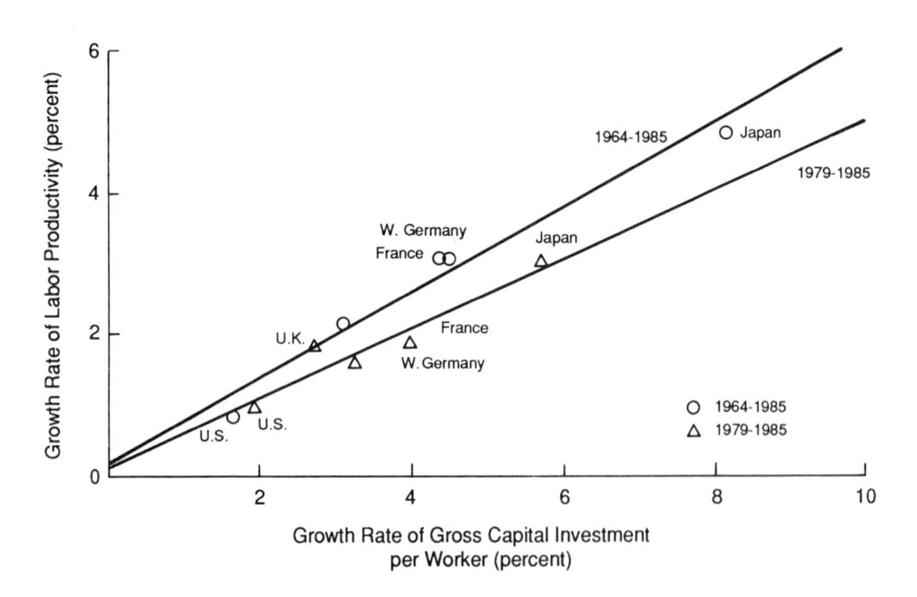

FIGURE 2 National productivity versus the capital-labor ratio. (Average annual percentage growth rates).

each country become critical for attaining higher sustained economic growth. It is an explanation for the difficulty the less developed countries have in exploiting generally available technology to accelerate their growth, and converging on the industrialized countries.

3. John Shoven's and Douglas Bernheim's (1989) work discloses that risk premiums matter to firms seeking investment, and these in turn depend on the "climate" set by governments and a nation's institutions. Some key climatic factors are (a) macroeconomic policies, (b) volatility and unpredictability of policy, (c) legal systems and institutions, (d) savings behavior, (e) educational systems, (f) science and technology policies, and (g) financial institutions.

In macroeconomic policy, differences in tax regimes and nominal interest rates between the United States and Japan are accompanied by different risks. Japanese companies appear to have a lower cost of capital (1/2 to 1/3 of the United States) across a wide range of investments, as recently confirmed by a major study of the Federal Reserve Bank of New York (McCauley and Zimmer, 1989). The results conform reasonably well to the relative industrial hurdle rates for the two countries. Better able to bear the risks involved, other nations invest for the longer term—and are increasingly forging ahead of the United States in key industries like electronics, machine tools, steel, autos, and the like. Moreover, as Ken-Ichi Imai (1989) of Hitotsubashi University has shown, Japan has a much better feedback loop than the United States between firms, government, and financial institutions, perhaps in part due to the number of engineers and technical experts in key policy roles, and the general lack of adversarial relations between business and government. Japanese firms are also able to establish wide networks of contractors and subcontractors, and some risks appear to be spread across the whole economy. In the United States, however, antitrust policy has, until recently, prevented such networks, and still limits cooperation to achieve results with potentially large social returns.

4. Good macroeconomic stabilization policy is a major boon to economic growth, report Ronald McKinnon and David Robinson (1989), as illustrated by the Japanese success in simultaneously reducing volatility after the oil shocks of the 1970s and maintaining rapid growth. The American policy of continual but uncertain dollar devaluation in the 1970s and 1980s is linked to higher inflation coupled with higher and more volatile nominal interest rates in the United States. These higher nominal interest rates reduce the value of tax depreciation allowances, raising the cost of capital, and increase the amount of borrowing necessary for financing the purchase of relatively longer-lived assets. Thus, the expectations of greater uncertainties have further

shortened the time horizons of American firms. In Japan, saving and capital investment were and are strongly favored, and productivity growth was very high, with low inflation. Policies favoring stabilization and capital formation support higher levels of economic well-being in the long as well as the short run. Durlauf's results by a quite different methodology lead to the same conclusion. The rapidity of technical change is obliterating the distinctions between short and long run.

THE LEGAL CLIMATE

Space permits only a brief mention of another important climate for growth—that of the legal system. It is a fact that the American legal system is unique among industrial nations in size, function, and complexity. With over 700,000 lawyers vs less than one-tenth as many in Japan (not all that different from the European countries), the contrast is stark, as are the effects. As Peter Huber (1989) said in a September 1989 conference sponsored by the Program on Technology and Economic Growth in Stanford's Center for Economic Policy Research, "Alone in the world so far, U.S. courts have abandoned the negligence standard for product liability; they ask juries to pass judgment instead on the adequacy of product design and manufacture . . . under a standard of strict liability for product defects; however, the people themselves, and their good care, good training, and good faith, are irrelevant. The new inquest concerns the product itself and its alleged defects. Today's U.S. tort system places technology itself in the dock." Huber cites many other changes in court decisions which are not happy ones for the innovator or for American competitiveness, such as excessive strictness on safety warnings; use of improved later designs to impeach the earlier designs if an accident or injury resulted therefrom; great latitude in filing of suits long after the machine or product was designed and used; and the rise of punitive damage awards. It is possible to identify various U.S. industries that have slowed down or reduced their commercialization of products because of liability uncertainties—general aviation, contraceptives, medical equipment, new drugs, vaccines, sporting goods, pesticides, etc.

Huber's conclusion says: "In the end, the search must be for rules that allow society to say yes to new and better products [and processes], with the same conviction and force that an open-ended liability system can say no to old and inferior ones. In many areas of policy, the answers given depend largely on the questions asked. For several decades, U.S. policymakers in the courts and elsewhere have asked:

What is unduly risky? And how can risk be deterred? But an equally important pair of questions is: What is acceptably safe? And how can safety be embraced?" The answers are not yet clear, and this uncertainty contributes significantly to the problems of competitiveness and growth.

PHYSICAL CAPITAL FORMATION

Our own and a number of other cross-sectional growth studies (such as by Philip Turner [1988] of the Organization for Economic Cooperation and Development, and John Helliwell and Alan Chung [1990] of British Columbia, Lawrence Summers [1990] of Harvard University, William Baumol and colleagues [1989] of New York University) over the last 25 years or so, establish beyond a reasonable doubt that there is a close correspondence between capital investment per worker and growth rates. There has been substantial disagreement between economists, however, about the direction of causation. In one sense, it does not really matter; the correlation itself indicates that neither demand nor supply of goods and services can be neglected. Indeed, it is well agreed that long-term growth takes place in the microeconomy—the true supply-side economics. Stabilization of demand by macroeconomic control of fiscal and monetary policy is traditionally used for short-term cyclical effects. Evidently, then, over some decades, as Solow and others have pointed out, an investment boom for the United States would be very beneficial to the long-run welfare of the American people. Short and long run really coincide, as Durlauf says.

In the most realistic sense, however, one should not expect that supply can be turned on and off as rapidly as macroeconomic policy (often a matter only of months in the latter). The gestation period for physical investment is 3-5 years; for R&D, perhaps 5-10; and for education and worker training, up to one generation. Diffusion of technology may take several years or even as long as the seventeen year life of a patent. Thus, if the close match of capital per worker and productivity is to persist, policies for long-term investment are, as summarized above, required on a concerted basis for many years. The business cycle and growth must be viewed from one overall perspective. These considerations are entirely consistent with the observations made by Barry Bosworth (1989) of the Brookings Institution, who estimates real rates of return as averaging about 8 percent for physical capital, 10 percent for education (with which our findings on the value of one year additional schooling concur), and perhaps in excess of 15 percent for research and development. (This is consistent with findings by Bernstein and Nadiri [1988] as well as Edwin Mansfield [1986] of the University of Pennsylvania. Furthermore, these researchers measured a much greater social return for R&D). For this reason, Bosworth likewise recommends increased capital formation of all kinds.

It would, therefore, be naive to assume an instantaneous match between supply and demand. Thus, causation is really irrelevant in practice, even though George Hatsopoulos of the Boston Federal Reserve and Thermo Electron Corporation, Summers, and Krugman (1988) believe it runs from investment per worker to growth. They just have to go together, and good policy must see to that. Now that the increasingly open nature of the world economy seems irreversible, there are increasing restraints upon a country's ability to manage demand stimulation by macroeconomic policy. The world capital markets impose their discipline on national governments. Demand management will increasingly seem to be synonymous with a stable pro-growth, pro-investment policy.

The basic conclusion from this and other recent research is that the role of physical capital matters very much indeed for the growth process, as does the proper and stable management of macroeconomic policy by government and the effective direction of many individual firms in the private sector. The United States has not been doing well in this area, the results demonstrate. High cost of capital, greater expectations of uncertainty, inadequate savings and investment, increasing reliance on foreign capital in an open world capital market, a tax system that is biased against saving and investment, short term horizons by managements and governments—these are all indicia of what Charles Schultze of the Brookings Institution calls the "termites" gnawing at the growth of the American standard of living. The Brookings volume on American Living Standards, Winter 1988-89, gives many details and analyses of our present position, as does a special section of *The Economist* of 24 September 1988. Perhaps the American people have the right to opt for more consumption now, and less investment for future consumption, but they should at least be made aware of the consequences of such a choice.

Despite the substantial consensus on these perceptions reached at a large Washington conference on "Saving—The Challenge for the U.S. Economy," organized by the American Council for Capital Formation in October 1989, and the general agreement by many political figures, there appears to be no political consensus at this time to establish a set of policies that would favor high investment and savings rates and be of greater predictability. These policies would propose stabilization of monetary and fiscal policy, so that investor expectations of inflation and uncertainty can change, and yield up to a 3 percent reduction in real interest rates. The tax code could be improved by correcting distortions and biases against savings and investment, particularly the latter. And investment is best in a low inflation environment.

R&D AND EDUCATION

What of the other two legs of the stool of growth? Intangible capital is equity capital, the most expensive kind, because it is generally not financed by borrowing. There are various estimates of the cost of equity capital, derived from stock market and tax considerations, but it is clear that it is substantially higher than long-term interest rates, as shown in Figure 3. The higher cost of equity (in the later 1980s about two and a half to three times as high as the real long-term after-tax cost of debt and more than twice the Japanese real after tax cost of equity, as disclosed by Hatsopoulos at the April NAE symposium; see Hatsopoulos, this volume) defines the rate at which

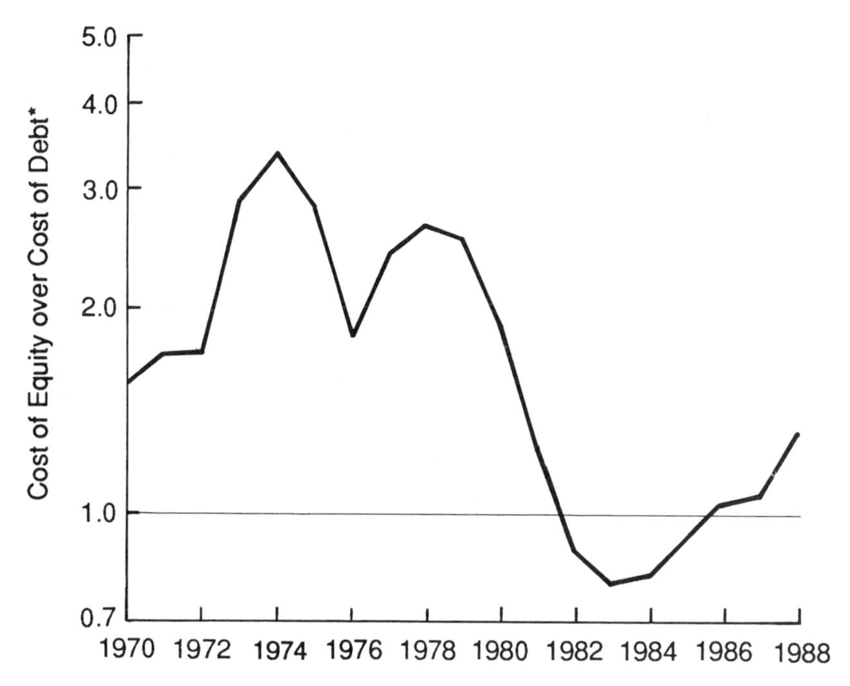

FIGURE 3 Nominal cost of equity. * Interest rate on 10-year Treasury notes. SOURCE: Hatsopoulos, George. 1989. Paper presented at American Council on Capital Formation conference "Saving—The Challenge for the U.S. Economy," 11-13 October.

future benefits from technology are discounted. (Of course, the marginal cost of capital will vary from company to company and may deviate significantly from the national average.) This fact, combined with the high real interest rates prevalent in the United States (required to ration the low domestic savings rate and attract foreign capital), as shown in Figure 4, has tended to steadily reduce investments by U.S. firms, and impairs the ability of American companies to compete on a long-term basis with the Japanese and Germans, among others. These countries have essentially the same overall average level of technology as the United States, although the level in individual industries may differ, so that the slope of the lower curve in Figure 2 indicates the importance of capital investment (interacting with technology) for growth. The data of Figure 2 generally conform to the findings of Boskin and Lau.

As to R&D expenditure trends, it seems as a result of the foregoing considerations, that from a level of 12.7 percent annual growth from 1976 to 1985, the present rate of increase has declined to about 6 percent in nominal terms. In real terms, it is a fall from 6.6 percent to 1.8 percent. In 1989 this has been the third time in four years that R&D investment has expanded more slowly than the economy itself. The United States must increase its R&D expenditures in both the public and private sectors if it is to maintain a high growth strategy.

The subject of educational deficiencies in the United States is too well known to warrant detailed discussion, but is included in the following section on microeconomic considerations. The striking effect of labor quality has already been shown.

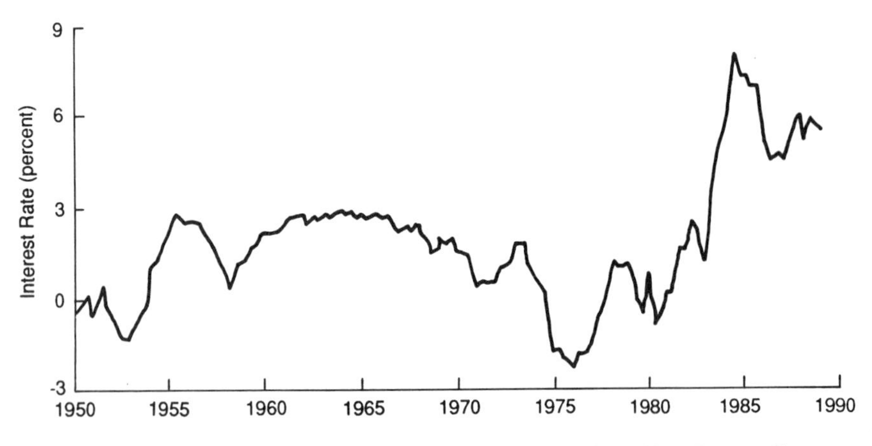

FIGURE 4 Long-term real interest rates. SOURCE: Neal Soss, First Boston Corporation.

MICROECONOMIC CONSIDERATIONS

There were many other research results reported at the Stanford Conference, where the foregoing papers were presented. Within an appropriate climate set by government, business firms are responsible and accountable for implementing the national goals for growth and competitiveness, and they are not the maximizing agents of macroeconomic analysis, as mentioned above. The following is a brief summary of what the speakers disclosed about how firms can improve their performance.

1. Conduct of research, particularly in basic science, is not America's most pressing problem. The problem rather is in the subsequent exploitation of the new products that emerge from research, especially in engineering, which is what leads to economic benefits. The Japanese have shown that, drawing on the generally available scientific base developed in all countries (particularly in the United States), they can advance their economic growth tremendously by engineering and commercialization techniques of their own creation. Different firms have very different internal capabilities (perhaps partly as a result of their own separate histories and the path dependencies that go along with those histories) and they also have access to very different bodies of technological knowledge, some of which are proprietary. More can certainly be done by firms to increase their access to richer bodies of data and knowledge.
2. Technology can be found in many places outside a particular firm, and there are many methods available to tap this, such as licensing, joint ventures, consortia, and contracting with universities. In-house research needs to explore avenues not available elsewhere, and this is what lies behind the many European consortia such as Esprit, Europa, and Jessi. The Japanese have proved to be masters of this process, the United States less so.
3. Innovation has often been thought of only in its technological terms, and not in economic terms. Thus, we commonly think of innovation in terms of entirely new products or components—transistors, television, computers, and petrochemicals—and much less in terms of the perhaps equally important subsequent cost reductions or performance improvements. This type of technical innovation has long been a traditional American strength. Indeed new products that are in no way major innovations in the sense of constituting drastic departures from the past are often highly profitable when they are correctly commercialized.
4. In the very important but often overlooked area of incremen-

tal innovation, some U.S. industries and firms have not yet mastered the skills necessary for shorter product cycles and rapid cost reductions (what Ralph Gomory [1989, 1990], formerly of IBM, contrasted with "ladder" or breakthrough innovation). He labels the one at least as important a factor in competitiveness of firms as the other. Industries and firms indeed vary widely, and some have managed the necessary transitions very well. By contrast, the Japanese have a less flexible research system but a much greater ability to design for manufacturing and short cycle times.

5. Engineering skills, therefore, need to be developed that conform more effectively to the requirements of the competitive process as described above: speed of adjustment to changes in market demand, shortening of product cycles, and greater attention to quality improvement and reliability. Manufacturing engineering education in the United States has not developed the overall systems design approach of the chemical engineering profession. That profession, in its linkage between the specific performance of individual pieces of equipment and their function in the overall plant embodying them, is a metaphor for the linkage of macroeconomics to its microfoundations, a linkage that is lacking in the bulk of current economic modeling. Most important, greater interrelations (feedback loops) must be established between users and suppliers, R&D marketing and manufacturing, and between physical products, software, and services.

6. Managements need to become much more skilled in using technology for greater competitiveness in a global marketplace. A longer-term view is essential, but this is tempered by the prevailing macroeconomic climate in the United States that urgently needs to be addressed. Firms will be understandably reluctant to spend money for innovative activity unless they have some reasonable degree of confidence that they can draw adequate financial benefits from the findings of R&D.

7. By moving into other countries (transplantation), including especially the United States, to gain global market share, the Japanese may well acquire some of the American skills in entrepreneurial R&D and innovation, although their efforts to adapt to other cultures have not always been successful. There is a general movement by firms in the major industrial centers to spread into the other major markets, not only to be nearer their customers, but also to hedge their exposure to the variability of national policies. Yet it is still largely true that design, engineering, and proprietary knowledge are concentrated in the home country. Even where an industry has been consistently successful, as in chemicals, it has still been unable, because of the problems of the American business climate, to capitalize fully

on its strengths by moving abroad as aggressively as its European competitors who are now buying up American chemical companies. Such foreign acquisitions of other American high-technology firms are occurring with increasing frequency and publicity.

8. The secondary educational system of the United States is inferior to that of the Japanese, but its university system is more effective and links more efficiently with industry. U.S. government science and technology policy has been constructive in this regard, although much money is spent on large projects that have no clear link to commercial products and services. Furthermore, Japanese companies spend a greater percentage of their GNP on civilian R&D than does the United States.

9. The American governmental system is less well equipped to deal with long-range strategy or to address difficult decisions than that of the Japanese, absent a crisis. In turn, many firms have similar difficulties, and the high cost of capital in addition to the methods of financial analysis combine to focus attention on short-run investments, of which takeovers and leveraged buyouts are symptoms. The Japanese "patient money" approach is the antithesis of this American situation. It is because of this imperative of the American innovation process that the viewpoint of the chief executive officer of a firm often seems so perverse and incomprehensible to the technical and research staff, and conversely. Yet, it is a prime reason why so many firms have lawyers and MBAs for chief executive officers, unlike the Japanese where technologists predominate. Indeed, as compared with the situation at the end of World War II, the cream of American college graduates no longer opts strongly for careers in science and engineering. The literate elite gravitate more to the law, the numerate to finance and business. This is not surprising; the financial incentives are much greater in these professions. If managements are to be serious about competing in the world of the 1990s, they must raise the rewards for young people to go into science and engineering, such as manufacturing engineering. It may also be beneficial to offer better subsidies to encourage students to go into advanced engineering training. The Massachusetts Institute of Technology has pioneered in this respect through its funding of its internationally known School of Chemical Engineering Practice, and its Leaders for Manufacturing program. More efforts of this nature can prove to be very important to the economy. The federal government could assist by proposing grants to young people who complete four years of college majoring in science or engineering and further grants for completing a Ph.D. in such fields.

10. The financial system of the United States is a major contribu-

tor to a higher cost of capital, because managements of large publicly held companies that own very little of their company's stock are fiduciaries for institutional investors who now constitute about half (and in many cases much more) of the owners. Accordingly, managers become increasingly risk averse and short term minded, the higher the cost of capital, because the separated owners have no real way to measure performance except on a carefully monitored financial performance basis. Furthermore, these institutional investors are themselves managed by fiduciaries for money contributed by large numbers of workers to their pension funds, and similar as well as additional constraints apply to them. *Management of fiduciaries by fiduciaries*—a system totally in contrast to the much closer interrelationship between owners and managers in Japan and Germany, and in the better privately held or owner-managed companies in the United States. The successful managerial capitalism of the first half of this century has been replaced by a very different form in both substance and style. Stiglitz has recently studied the disturbing effects of the capital markets on productivity growth, with somewhat similar conclusions to those contained in this paper (Greenwald et al., 1990). The 20-page insert on capitalism in *The Economist* of 5 May 1990 deals with this issue also.

Of course, financial institutions respond to clear evidence that a firm's management is effective by awarding it a higher price-earnings multiple than the market average, which lowers its cost of equity capital. If on the other hand, their conclusion is the reverse, the temptation for takeovers, leveraged buyouts, mergers, and acquisitions becomes great when aided by a tax system that favors debt over equity. Managements often fail to inform their investors adequately about their actions, but most are only too well aware of the financial constraints.

WHAT SHOULD BE THE NATIONAL GOAL

Our most important finding, as described above, has been that investment in human and nonhuman capital accounts for the largest part of U.S. economic growth during the postwar period. The slowdown since 1973 has resulted in a full percentage point lower growth rate relative to the preceding postwar average. The need for new pro-investment policies of all three kinds is best illustrated by consideration of the importance of accelerated growth in real income per worker for the proper funding of the Social Security System. Alicia Munnell (1989) of the Boston Federal Reserve, in analyzing a Brookings study (by Aaron et al. [1988]), pointed out that the *intermediate* projection of the Social Security Trustees suggests that the net real wage per worker in 2020 will be from 199 to 211 percent of today's level. Thirty years is not long term. How, then, is this going to materialize in view of the fact that real wages per worker have barely improved since the early 1970s? Since demographic projections show a significant decline in the working population (it will fall to around 1 percent per year in the later 1990s from a peak in the 1970s of about 2.5 percent), it is evident that a doubling of the real net wage can occur only by a massive increase in the rate of all kinds of capital investment per worker—physical, intangible, and human. Furthermore, we must be able to pay the costs of the large amount of capital we have been importing from abroad.

This means raising the noninflationary annual growth rate (real GDP) from its current level to perhaps 3.5 percent. As we noted above, this is required for the compounding effect to produce a doubling of real living standards in a generation. Even though increasing capital investment can only gradually raise productivity, because of the enormous stock of existing capital, it can nevertheless be launched within roughly a decade, if the new capital investment is *efficiently* concentrated in the leading-edge industries that perform most of the R&D and thus affect most strongly the overall productivity of the economy. That this criterion is essential can be seen from the failures of the socialist countries to reap productivity benefits from their very high capital investments, which the market economies have managed more effectively. Summers has made some rough calculations that suggest that this improvement could be perhaps 0.5 percentage points a year (Boskin, 1988). The 1990 Economic Report of the President makes similar calculations, and points out that this signifies a major long-term improvement in living standards, as we have also stated at the beginning. However, it is important not to repeat the errors of the 1970s, when so much capital was funneled into relatively unproductive real estate and other investments.

HOW GROWTH RATES CAN BE INCREASED TO MEET THE NATIONAL GOAL

Thus, taking these considerations into account, Boskin has described the relation between Solow's general framework and the more recent results as follows. Under Solow, the fundamental variables that increase the rate of real per capita growth of a country in the long term are the rate of technical change and the increase in quality of the labor force. Increasing just the capital-to-labor ratio by this theory will lead only to a *temporary* increase in the rate of growth (moving from growth path 1 to growth path 2 in Figure 5), but to higher living standards—a desirable goal in itself. As mentioned, such an increase in physical capital formation occurred during the 1960-1979 period in Japan. These large growth rates proved difficult to sustain for most of the 1980s, but permanent advantages for many industries and for the population have been created.

Measurements of productivity growth alone are not, however, a complete expression of the role of technology in economic growth. As mentioned above, in the original formulations, and in much of the work that followed, the inputs of labor, capital, and productivity were deemed essentially independent of each other. The contrary findings of Boskin and Lau can perhaps be more easily interpreted from our actual experience in the innovative process. R&D and creative design are seldom performed all by themselves—but rather only when they are expected to be employed in new or improved facilities or in superior operating modes. So technological change is not only embodied in physical capital investment, it is itself capital—intangible capital—and also a powerful inducement to it, since the availability

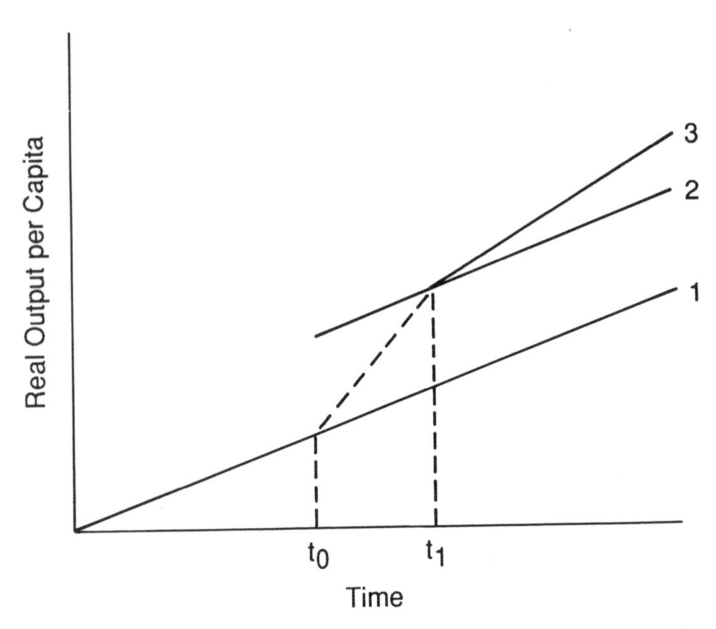

FIGURE 5 Alternative growth paths: Technical change and capital formation. t_0: proinvestment policy leads to higher capital formation and transition to higher level of income. t_1: economy resumes long-run growth rate or through interaction of investment and technical change, moves to more rapid growth path. SOURCE: Boskin (1986, p. 37).

of superior technology is a major incentive to invest. Investment in turn stimulates more R&D and creative learning. Likewise, improvements in labor quality (human knowledge, skill, and training) are both a requirement of and a spur to technological change and are another form of investment—human capital. Capital investment creates a favorable emotional and intellectual commitment to people and evokes individual creativity. The building of a new facility with improved technology by one company puts pressure on its competitors to do likewise.

Thus, unlike the inputs of labor and capital to the growth process, which are more direct, the contribution of technology has a multiplier effect—an externality. All three combined produce greater growth. As we have discussed in our recent book (Landau, 1988, 1989b, 1990a), technology now often takes an embodied and reinforcing form within each of the basic factors of production—labor and capital—to a far greater extent than was thought before. There are multiple feedback loops. And when workers, managers, and technologists use such capital investment, particularly when they feel a sense of participation, they are also learning from and drawing upon an expanded store of human knowledge, which yields continuing improvements in efficiency and output. The recent report of the Massachusetts Institute of Technology Productivity Commission (Dertouzos et al., 1989) makes this same point.

Thus, Boskin's growth path 3 in Figure 5 shows that with these interactions between technological change, physical capital, and labor quality, a higher rate of capital investment can move the economy to a higher rate of medium-term economic growth, as well as lead to an upward shift in the level at any given time. He believes, therefore, and we concur, that the rate of investment and technical change are positively linked. The traditional inputs to the production process are not, in fact, independent variables. This is especially true when it comes to exploiting the results of "breakthrough" R&D, which require large new investments. This increased growth path may be viewed as a series of transitions in a dynamic economy never really at equilibrium or at a steady state because of continuing unpredictable, endogenous technical changes. If technical change is not exogenous, embodiment and learning by doing (phenomena that the neoclassical growth economists did recognize) interact with capital investment to improve growth rates, and capital investment is critical in reaching a higher equilibrium and approaching the technological frontier throughout the economy at a faster rate. This is the evolutionary process by which the nonconvexity of the production process is established, and unceasing growth occurs. Hence, in view of the substantial number of really novel technologies now available and the effects of continuing R&D and design efforts, the need for totally new facilities and closing down of obsolete units is becoming much greater—a version of "catchup" for the United States, particularly in some of its industries.

CONCLUSION

There seems to be little remaining doubt among economists that to improve American living standards and maintain American influence in international affairs, increased investment in all kinds of capital per worker will be necessary, especially considering the growing environmental concerns. The cost of funds (the mix of debt and equity) is a fundamental driving force in the private sector decisions that lead to such accumulation. However, the basic consideration for physical capital investment is the cost of capital, which is the pretax return required to pay all taxes and depreciation on plant and equipment. Public investments in infrastructure are also important, and are certainly sensitive to the savings rates. With declining demographic increases in the work force, enhanced productivity improvement obtained in this way need not be at the expense of job creation. As Table 3 demonstrates, the policies of the United States in the past two decades have had a very beneficial effect on job formation (unemployment is now 5.3 percent) compared with Europe and Japan, where productivity and capital formation were higher, but so was unemployment (over 8 percent in Europe). A large measure of this accomplishment

TABLE 3 Employment-Civilian Millions

Year	EEC	USA	Japan
1955	101.4E	62.2	41.9
1965	104.8E	71.1	47.3
1975	105.5	85.8	52.2
1985	106.7	107.2	58.1
1986	107.5	109.6	58.5
1987	108.3	112.4	59.1
1988	110.8	115.0	60.1
Net Increase	9.4	52.8	18.2

NOTE: EEC = The Ten; E = estimate.

SOURCE: OECD, EEC, Bank of Japan, IFO. Courtesy of *The Economist*.

comes from the American entrepreneurial ability to generate many new small and medium sized companies. The price we paid, as shown in Figure 4, was a rise in long term real interest rates, and an inevitably weak capital formation (Figure 6). Figure 7 shows the growth rates per employed worker to emphasize this divergence in national results.

Business investment, of course, is cyclical. The low (often negative) real interest rates of the 1970s led to a capital spending boom and the overcapacity in some industries cited above, with low productivity gains. The rise in rates in the 1980s has shortened the horizon of investors. Nevertheless, it must be reiterated that interest rates are only one component of the cost of capital to firms, which depend in varying degree on mixes of debt and equity. The United States must get its cost of capital down to the level of its international competitors by, among other structural measures, removing the tax biases against productive investment and savings, so that it is no longer necessary to pay a high premium to import foreign capital. As Durlauf says, growth is a function of how each economy is managed, despite the internationalization of capital markets.

This paper is not the place to enlarge on this theme, but it will clearly necessitate all of the goals described above. Such an altered policy is not now in place. Because, in a scarce savings economy (and domestic savings and investment are still linked in an open economy, although less tightly), returns to financial assets in the 1980s exceeded returns to many physical assets in the real sector, the economy had to adjust by rationalizing the use of capital so that it could compete with the returns available on its paper image. This adjustment is in the form of a lower investment pattern and entailed mammoth equity retirements, mergers and acquisitions, leveraged buyouts, privatization, and "junk" bonds. Restructuring responds to the need to make physical capital productive enough to withstand the high real interest rates required by the financial markets. However, the cost may well be greater vulnerability because of the greater indebtedness, and the erosion of the critical but capital-intensive manufacturing base. The net interest payments of nonfinancial corporations rose from 8.6 percent of cash flow in late 1959 to 24.2 percent by the late 1970s; in 1989 year this figure reached almost 26 percent (and exceeded after-tax profits by $40 billion). Bankruptcies have been increasing. The savings and loan crisis is a glaring example.

In significant and sobering contrast is the current performance of the Japanese economy. As the *Wall Street Journal* has recently reported (Ono, 1989), Japanese capital spending has been growing at double-digit rates. Many industries, from autos to computer chips to shipbuilding

FIGURE 6 Weak capital formation in the 1980s. SOURCE: Datastream International/ Worldview.

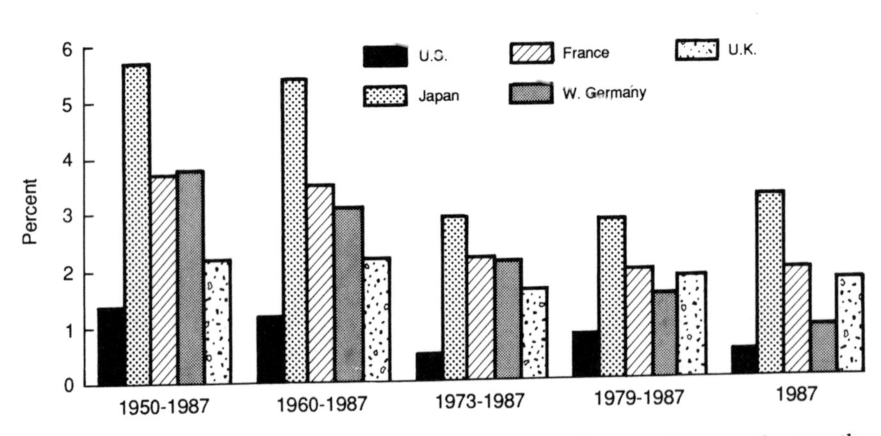

FIGURE 7 International comparison of average annual rates of economic growth. Average annual percent change in real gross domestic product per employed person. SOURCE: U.S. Department of Labor.

are modernizing, expanding capacity, and developing new products. In the 12 months ending 31 March 1989, Japan's capital outlays exceeded those of the United States (measured at prevailing exchange rates), $521.4 billion compared with $494.8 billion, despite the fact that Japan's GNP is less than two-thirds that of the United States. It was probably well over $600 billion for Japan in all of 1989 compared with somewhat more than $500 billion for the United States. Figure 6 shows the divergence between the two countries. Much of this investment boom is in the high-tech industries that already threaten the viability of many American firms. It seems increasingly clear that the United States now has a new economic rival, as the Soviet Union retreats, and one far more formidable because it is so difficult to develop a "crisis" mentality to spur Americans to change their habits.

This Japanese investment boom is fueled by an enormous capital market that channels the large Japanese private savings into productive enterprises. Debt capital is available at interest rates still below those of the United States. Equity issues are far larger than in the United States. Whereas in 1986, American companies raised a record $67.9 billion from the sale of stock and similar securities, by the first eight months of 1989, Japanese corporations raised more than $110 billion, compared with $20 billion by American companies. Can anyone doubt that in a relatively few more years, Japanese productivity advances will greatly exceed ours, and new plants will be able to supply world markets with even more and better products? A recent visit to Japan confirms that there are many other favorable factors to fuel rising Japanese competitiveness: a homogeneous disciplined population; an excellent secondary educational system; a manipulated financial system; a tax system favoring saving over consumption; a strong governing party (even if somewhat weakened); a strong and competent bureaucracy that favors its industry in domestic and foreign markets; a relatively mild antitrust stance; and fierce competitiveness between firms, among many special characteristics. Perhaps Japan has even improved upon the capitalist system that we invented along with the British! But it has not achieved perfection. The recent fall in the Japanese stock market and rising inflation bring to an end the speculative bubble in stock and land prices, raise somewhat the cost of Japanese equity, and herald a reduction in the flood of capital flowing overseas, especially to the United States. We may, therefore, see continuing high real interest rates and slow growth here, unless our national savings rate improves, as it has recently begun to do. The adaptable Japanese system, however, continues to hold many advantages, and Japan should continue to grow over the longer term at a higher rate than other industrialized countries.

There is a massive job ahead to change the direction of our economy. The American people need to understand the choices they must address. These must link the three types of capital (physical, human, and intangible) to the current economic situation. Our argument for more capital investment is grounded in the economic and technical opportunities facing the United States now. The current importance of the great (and as yet only very partially realized) information technologies revolution means that technical progress is embodied in physical capital to an unprecedented extent. The same is true for the new materials and biotechnology. This embodiment links investment in new knowledge for growth with investment in new physical capital. Similarly, opportunities for technical advance are linked with investment in R&D and human capital. These new technologies require a different skill mix in the work force, especially at the high skill end, where the American higher education system offers a source of substantial comparative advantage to the United States, if exploited with new investment in physical and intangible capital. This is also an imperative because the major Japanese exports to the United States are both research intensive and capital intensive, and to compete, firms need all three legs of the stool of growth. Yet, current policies *disfavor* investment in general, and indeed have tended to encourage investment in the less productive forms of capital.

How can these choices be made constructively in the face of profound skepticism about the efficiency and scandals of government? As Michael Porter (1990a,b) of the Harvard Business School says, direct intervention by the Japanese government has been abandoned, and this is appropriate at Japan's advanced stage of competitive development. America should not be pushed into adopting policies that do not work in an advanced economy, and instead should concentrate on getting the overall climate right. Nevertheless, many attempts are being made by various interests and scholars to justify managed trade, and by some technologists and businessmen who feel unable to compete against foreign-managed trade and buyouts of critical technologies and urge government counterefforts.

A more reasoned approach lies in the question of where government can intervene *effectively* in the microeconomy. In March 1990 President Bush committed his administration to fostering critical precompetitive generic technologies that "support both our economic competitiveness and our national security." Robert M. White (1990), president of the National Academy of Engineering, in a speech in April 1990 developed this theme further in a thoughtful way, but there is as yet no national consensus on where generic technology lies between basic research and applied industrial development, and to what extent and by what means it should be addressed.

However, the President's reference to national security reminds us that there are many who feel that certain industries (particularly semiconductors) are essential to national economic security, and that both America's inadequate macroeconomic policies and the failings of American firms and industry structures are going to hand control of critical technologies and companies to foreign-based companies, especially the Japanese. They are encouraged by the recent growth and trade theory developments (referred to earlier) that suggest some protection in early stage technology may be positive for economic growth. It is not clear, nevertheless, whether U.S.-owned companies behave differently than foreign-owned businesses in the United States, or whether control of technology is associated with national ownership. Companies are increasingly becoming not just multinational but global—perhaps slipping beyond the control of *any* national government. So far, however, this has been less true of the Japanese, and this gives rise to strong protectionist feelings on the one hand, and surrender of hope for American firms' ability to compete on the other, and hence a desire to see more aggressive firms from abroad locate in the United States, even at the expense of American-owned firms.

Considering these contentious forces, the desire for greater national controls and strategy versus the spread of supranatural global companies based in many countries, it is no wonder that there is a growing concern among many economists that the United States will therefore be increasingly pushed toward protectionism of various kinds. Their concern is over costly intervention in the private sector, such as the proposal by a National Advisory Commission on Semiconductors for government funding of a venture capital corporation at a multibillion-dollar level to resurrect the defunct American consumer electronics industry, despite the well-established inability of such policies to ensure success. In the absence of the necessary industrial structure, reviving a whole industry would have ramifications throughout many other sectors of the economy and incur huge costs. Considering how hard such intervention can be, the best that any government can do in its four-year term is probably to focus primarily on getting the macroeconomic policies right as its highest priority.

When one couples this fundamental problem with the growing and frequently legitimate pressures of the environmental advocates, and the deteriorating infrastructure, which also requires large investment (and which is conducive to productivity gains in the whole economy [Munnell, 1990]), there must arise a gnawing fear that America's position in the world may slip before too many years into a second-class role, and that growth in living standards will be inadequate to address the many social problems and inequities that exist today. The proper remedy, as pointed out in this paper, is becoming quite clear, but it will require patience, determination, and leadership, a change in fundamental perceptions of national priorities, and abandonment of obsolete economic theories and perspectives.

NOTES

1. In this article, the use of "we" implies references not only to my work, but to other work (or experience) done (or had) at Stanford and Harvard universities. I am particularly indebted to the directors of the programs on technology and growth, Nathan Rosenberg and Lawrence Lau (Stanford) and Dale Jorgenson (Harvard), with whom I serve as codirector of both programs. Thanks are also due to Paul Romer of Chicago and Timothy Bresnahan, Steven Durlauf and John Shoven of Stanford. However all errors are my sole responsibility.
2. As a consequence of the current account deficit; in addition, central bank transactions may have resulted in perhaps another $50 billion inflow.
3. See Denison (1957, 1962, 1967, 1972, 1979, 1985).
4. See Griliches (1979, 1988), Griliches and Jorgenson (1967, 1972a, 1972b).
5. See Jorgenson, Gollop, and Fraumeni (1987); Jorgenson, Kuroda, and Nishimizu (1986); and Jorgenson (1988).
6. See Kendrick (1961, 1973, 1976, 1983) and Kendrick and Grossman (1980).

REFERENCES

Aaron, H., B. Bosworth, and G. Burtless. 1988. Can America Afford to Grow Old? Washington D.C.: The Brookings Institution.

Abramovitz, M. 1956. American Economic Review 46(May):5-23.

Arrow, K. 1962. The economic implications of learning by doing. Review of Economic Studies 29(June):155-173.

Baumol, W. J., S. A. Baley Blackman, and E. N. Wolff. 1989. Productivity and American Leadership: The Long View. Cambridge, Mass.: MIT Press,

Bernstein, J. I., and M. I. Nadiri. 1988. Interindustry spillovers, rates of return, and production in high tech industries. American Economic Review (May): 429-434.

Boskin, Michael J. 1986. Macroeconomics, technology, and economic growth: An introduction to some important issues. Pp. 33-56 in The Positive Sum Strategy: Harnessing Technology for Economic Growth, R. Landau and N. Rosenberg, eds. Washington, D.C.: National Academy Press.

Boskin, M. J. 1988. Tax policy and economic growth: Lessons from the 1980s, Journal of Economic Perspectives 2(4)(Fall):87

Boskin, M., and L. Lau. 1989. Capital Formation and Productivity Growth: An International Comparison. Paper presented at Conference on Economic Growth and the Commercialization of New Technologies, Center for Economic Policy Research, Stanford University, September.

Boskin, M. J., and L. J. Lau. 1990. Post-War Economic Growth in the Group-of-Five Countries: A New Analysis. Working paper. Department of Economics, Stanford University, July.

Bosworth, B. Testimony before the Committee on Ways and Means, House of Representatives. United States Congress April 19, 1989.

Christensen, L. R., and D. W. Jorgenson. 1969. The measurement of real capital input, 1929-1967. Review of Income and Wealth 15(4)(December):293-320.

Christensen, L. R., and D. W. Jorgenson. 1970. U.S. real product and real factor input, 1929-1967. Review of Income and Wealth. 16(1)(March):19-50.

Christensen, L. R., and D. W. Jorgenson. 1973. Measuring the performance of the private sector of the U.S. economy, 1929-1969. Pp. 233-351 in Measuring Economic and Social Performance, M. Moss, ed. New York: Columbia University Press.

David, P. A. 1990. The dynamo and the computer: An historical perspective on the modern productivity paradox. American Economic Review. 80(2)(May):355-361.

Denison, E. F. 1957. Theoretical aspects of quality change, capital consumption, and net capital formation. In Conference on Research in Income and Wealth, Problems of Capital Formation. Princeton, N.J.: Princeton University Press.

Denison, E. F. 1962. Sources of Economic Growth in the United States and the Alternatives Before Us. New York: Committee for Economic Development.

Denison, E. F. 1967. Why Growth Rates Differ. Washington, D.C.: The Brookings Institution.

Denison, E. F. 1972. Final Comments, Survey of Current Business, Part II. 52(5):95-110.

Denison, E. F. 1974. Accounting for United States Economic Growth, 1929-1969. Washington, D.C.: The Brookings Institution.

Denison, E. F. 1979. Accounting for Slower Economic Growth, Washington, D.C.: The Brookings Institution.

Denison, E. F. 1985. Trends in American Economic Growth, 1929-1982. Washington, D.C.: The Brookings Institution.

Dertouzos, M. L., R. K. Lester, R. M. Solow, and the MIT Commission on Industrial Productivity. 1989. Made in America: Regaining the Productive Edge. Cambridge, Mass.: MIT Press.

Durlauf, S. 1989. International Differences in Economic Fluctuations. Paper presented at Conference on Economic Growth and the Commercialization of New Technologies, Center for Economic Policy Research, Stanford University, September.

Eatwell, J., M. Milgate, and P. Newman, eds., 1987. New Palgrave. New York: MacMillan.

The Economist. September 23, 1989. Editorial pp. 13-14, 81.

Gomory, R. 1989. The Technology-Product Relationship: Early and Late Stages. Conference on Economic Growth and the Commercialization of New Technologies, Center for Economic Policy Research, Stanford University, (September).

Gomory, R. 1990. Of ladders, cycles and economic growth. Scientific American (June):140.

Greenwald, B. G., M. A. Salinger, and J. E. Stiglitz. 1990. Imperfect capital markets and productivity growth. National Bureau of Economic Research conference paper, April.

Griliches, Z. 1972a. Issues in Growth Accounting, A Reply to Edward F. Denison. Survey of Current Business 52(4) Part II, (May):65-94.

Griliches, Z. 1972b. Issues in Growth Accounting, Final Reply. Survey of Current Business 52(5) Part II, (May):111.

Griliches, Z. 1979. Issues in assessing the contribution of research and development to productivity growth. The Bell Journal of Economics 10(Spring):92-116.

Griliches, Z. 1988. Technology, Education and Productivity: Early Papers with Notes to Subsequent Literature. London: Basil Blackwell.

Griliches, Z., and D. W. Jorgenson. 1967. The explanation of productivity change. Review of Economic Studies 34(2)(99)(July): 249-280.

Grossman, G. M., and E. Helpman. 1990. Trade, innovation and growth. American Economic Review 80(2):86-91.

Hatsopoulos, G. N., P. R. Krugman and L. H. Summers. 1988. U.S. competitiveness: Beyond the trade deficit. Science 241(July):299-307.

Helliwell, J. F., and A. Chung. 1990. Aggregate productivity and growth in an international comparative setting. In International Productivity and Competitiveness, B. G. Hickman, ed. New York: Oxford University Press.

Huber, P. 1989. Liability and insurance problems in the Commercialization of new products. Paper presented at Conference on Economic Growth and the Commercialization of New Technologies, Center for Economic Policy Research, Stanford University, September.

Hulten, C. R., and F. C. Wykoff. 1981. The measurement of economic depreciation. In Depreciation, Inflation and the Taxation of Income from Capital, C. R. Hulten, ed. Washington, D.C.: Urban Institute Press.

Hulten, C. R., J. W. Robertson, and F. C. Wykoff. 1989. Energy, obsolescence, and the productivity slowdown. Pp. 225-258 in Technology and Capital Formation, D. Jorgenson and R. Landau, eds., Cambridge, Mass.: MIT Press.

Imai, K. 1989. The Japanese Pattern of Innovation and its Commercialization Process. Paper presented at Conference on Economic Growth and the Commercialization of New Technologies, Center for Economic Policy Research, Stanford University, September.

Jorgenson, D. W. 1988. Productivity and postwar U.S. economic growth. Journal Economic Perspectives 2(4)(Fall):23-41.

Jorgenson, D. W., and B. M. Fraumeni. 1990. Investment in education and U.S. economic growth. In The U.S. Savings Challenge, C. E. Walker, M. A. Bloomfield, and M. Thorning, eds. Boulder, Colo.: Westview Press.

Jorgenson, D. W., and R. Landau. 1989. Technology and Capital Formation. Cambridge, Mass.: MIT Press.

Jorgenson, D. W., M. Kuroda, and M. Nishimizu. 1986. Japan-U.S. industry-level productivity comparisons, 1960-1979. In productivity in the U.S. and Japan, C. R. Hulten and J. R. Norsworthy, eds. Chicago: University of Chicago Press,

Jorgenson, D. W., F. Gollop, and B. Fraumeni. 1987. Productivity and U.S. Economic Growth. Cambridge, Mass.: Harvard University Press.

Kendrick, J. W. 1961. Productivity Trends in the United States. Princeton, N.J.: Princeton University Press.

Kendrick, J. W. 1973. Postwar Productivity Trends in the United States, 1948-1969. New York: National Bureau of Economic Research.

Kendrick, J. W. 1976. The National Wealth of the United States. New York: Conference Board.

Kendrick, J. W. 1983. Interindustry Differences in Productivity Growth. Washington, D.C.: American Enterprise Institute.

Kendrick, J. W., and W. S. Grossman 1980. Productivity in the United States, Trends and Cycles. Baltimore, Md.: Johns Hopkins University Press.

Landau, R. 1988. U.S. economic growth. Scientific American 258(6)(June):44-52.

Landau, R. 1989a. The chemical engineer and the CPI: Reading the future from the past. Chemical Engineering Progress (September):25-39.

Landau, R. 1989b. Technology and capital formation. Pp. 485-505 in Technology and Capital Formation, D. Jorgenson and R. Landau, eds., Cambridge, Mass.: MIT Press.

Landau, R. 1990a. Capital investment, key to competitiveness and growth. Brookings Review (Summer).

Landau, R. 1990b. Chemical Engineering: Key to the Growth of the Chemical Process Industries. AIChE Symposium Series 86(274):9-39.

Lindbeck, A. 1983. Econ. Journal 93(March):13-34.

Lipsey, R. 1990. NBER Working paper No. 3293.

Lucas, R. E., Jr. 1988. On the mechanics of economic development. Journal of Monetary Economics 22:3-42.

Maddison, A. 1987. Growth and slowdown in advanced capitalist economies: Techniques of quantitative assessment. Journal of Economic Literature. 25(June):649-698.

Mansfield, E. 1986. Microeconomics of technological innovation. Pp. 307-325 in The Positive Sum Strategy, R. Landau and N. Rosenberg, eds., Washington, D.C.: National Academy Press.

McCauley, R. N., and S. A. Zimmer. 1989. Explaining international differences in the cost of capital. Federal Reserve Bank of New York Quarterly Review 14(2)(Summer).

McKinnon, R., and D. Robinson. 1989. Dollar devaluation, interest rate volatility, and the duration of investment. Paper presented at Conference on Economic Growth and the Commercialization of New Technologies, Center for Economic Policy Research, Stanford University, September.

Munnell, A. H. 1989. Social Security Surpluses: How Will They Be Used? Paper presented at American Council for Capital Formation conference "Saving—The Challenge for the U.S. Economy," Washington, D.C., October.

Munnell, A. H. 1990. Why has productivity growth declined? Productivity and Public Investment. New England Economic Review (Jan./Feb.):3-22.

Nelson, R. R. 1981. Research on productivity growth and productivity differences: Dead ends and new departures. Journal of Economic Literature. 19(September): 1029-1064.

Nelson, R. R., and S. G. Winter. 1982. An Evolutionary Theory of Economic Change. Cambridge, Mass.: Harvard University Press.

Ono, U. October 26, 1990. Capital Spending. Wall Street Journal.

Porter, M. 1990. Harvard Business Review, (May-June):190-192.

Porter, M. 1990. The Competitive Advantage of Nations, New York: The Free Press.

Romer, P. M. 1986. Increasing returns and long-run growth. Journal of Political Economy 94(October):1002-1037.

Romer, P. M. 1987a. Crazy explanations for the productivity slowdown. NBER Macroeconomics Annual. Stanley Fischer, ed., Cambridge, Mass.: MIT Press.

Romer, P. M. 1987b. Growth based on increasing returns due to specialization. American Economic Review 77(May):56-62.

Romer, P. M. 1989a. Capital Accumulation in the Theory of Long-run Growth, in Modern Business Cycle Theory, R. Barro, ed., Cambridge, Mass.: Harvard University Press.

Romer, P. M. 1989b. Measurement Error in Cross Country Data, Manuscript.

Romer, P. M. 1990. Endogenous technological change. Journal of Political Economy.

Rosenberg, N., and R. Landau. 1989. Successful Commercialization in the Chemical Process Industries. Paper presented at Conference on Economic Growth and the Commercialization of New Technologies, Center for Economic Policy Research, Stanford, University, September.

Shoven, J., and D. Bernheim. 1989. Comparison of the Cost of Capital in the U.S. and Japan: The Roles of Risk and Taxes, Paper presented at Conference on Economic Growth and the Commercialization of New Technologies, Center for Economic Policy Research, Stanford University, September.

Solow, R. 1956. Quarterly Econ. 70:65-94.

Solow, R. 1957. Review of Economics and Statitistics. 39:312-20.

Solow, R. 1987. Nobel Lecture.

Summers, L. 1989. What Is the Social Return of Capital Investment? Paper presented at Robert Solow's 65th birthday symposium, MIT, April.

Turner, P. 1988. Saving and investment, exchange rates, and international imbalances! A comparison of the U.S., Japan, and Germany. J. Japanese Int. Econ. 2(3)(September):259-265.

White, R. M. 1990. Technology policy in an interdependent world. Paper presented at American Association for the Advancement of Science, Washington, D.C., April 13, 1990.

Wykoff, F. C. 1989. Economic depreciation and the user cost of business-leased automobiles. Pp. 259-292 in Technology and Capital Formation, D. Jorgenson and R. Landau, eds. Cambridge, Mass: MIT Press.

Capital Formation and Economic Growth

Michael J. Boskin and Lawrence J. Lau

Enhanced capital, labor, and technical progress are the three principal sources of the economic growth of nations. Since the rate of growth of labor is constrained by the rate of growth of population, it is seldom, especially for industrialized countries, higher than two percent per annum, even with international migration. Consequently, the rate of growth of capital (physical and human) and technical progress have been found to account for a significant proportion of economic growth by a long line of distinguished economists: Abramovitz (1956), Denison (1962a,b; 1967), Griliches and Jorgenson,[1] Kendrick (1961, 1973), Kuznets (1965, 1966, 1971, 1973) and Solow (1957), to name only a few. For example, Jorgenson, Gollop, and Fraumeni (1987, p. 21) found that between 1948 and 1979, capital formation accounted for 46 percent of the economic growth of the United States, labor growth accounted for 31 percent, and technical progress accounted for 24 percent.

Most studies of the sources of economic growth, or growth accounting, are based on the concept of an aggregate production function:

$$(1)\quad Y_t = F(K_t, L_t, t)$$

where Y_t, K_t, and L_t are the quantities of real aggregate output, capital and labor respectively at time t and t is an index of chronological time.[2] The purpose of growth accounting is to determine from the empirical data how much of the change in real output between say $t = 0$ and $t = 1$ can be attributed to changes in the inputs, capital and labor, and technology, respectively. Taking natural logarithms of both sides of equation (1) and differentiating it totally with respect to t, we obtain:

$$(2) \quad \frac{d\ln Y_t}{dt} = \frac{\partial \ln F}{\partial \ln K_t}(K_t, L_t, t)\frac{d\ln K_t}{dt} + \frac{\partial \ln F}{\partial \ln L_t}(K_t, L_t, t)\frac{d\ln L_t}{dt} + \frac{\partial \ln F}{\partial t}(K_t, L_t, t)$$

where $\frac{d\ln Y_t}{dt}$, $\frac{d\ln K_t}{dt}$ and $\frac{d\ln L_t}{dt}$ are the instantaneous proportional rates of change of the quantities of real output, capital and labor respectively at time t; $\frac{\partial \ln F}{\partial \ln K_t}$ and $\frac{\partial \ln F}{\partial \ln L_t}$ are the elasticities of real output with respect to capital and labor respectively at time t and $\frac{\partial \ln F}{\partial t}$ is the instantaneous rate of technical progress, or equivalently, the rate of growth of output holding the inputs constant.

The first term on the right-hand side of equation (2) thus represents the contribution of the growth of capital to the growth of real output. Note that the contribution of capital depends on both the production elasticity of capital and the rate of growth of capital. If the rate of growth of capital is low, then the contribution of capital will be low even with a high production elasticity of capital. Similarly, the second term represents the contribution of the growth of labor and the third term represents the contribution of technical progress. Together, the three terms add up to the rate of growth of real aggregate output.

However, not every variable on the right-hand side of equation (2) can be directly observed. Only the rates of growth of real aggregate output, capital and labor can be observed. The elasticities of output with respect to capital and labor must be separately estimated, often requiring additional assumptions. Moreover, note that the instantaneous rate of technical progress, $\frac{\partial \ln F}{\partial t}(K_t, L_t, t)$, depends on K_t and L_t as well as t. To the extent that K_t and L_t change over time, the rate of technical progress over many periods cannot be simply cumulated from one period to the next, unless technical progress is neutral, in which case the instantaneous rate of technical progress, $\frac{\partial \ln F}{\partial t}$, is independent of capital and labor.

Thus, in general, in order to use equation (2) to measure technical progress over time, three basic hypotheses are maintained: constant returns to scale, neutrality of technical progress, and profit maximization with competitive output and factor markets.[3] Profit maximization with competitive markets allows the identification of the elasticities of output with respect to labor with the share of labor cost in total output. Constant returns to scale in production implies that the sum of the elasticities of output with respect to capital and labor is exactly unity, so that the elasticity of capital can be readily estimated as one minus the elasticity of labor when the latter is known. Neutrality of technical progress justifies the cumulation of successive estimates of technical progress over time.

In a recent study, we attempt to identify and estimate the degree and bias of scale economies, the rate and bias of technical progress, and the elasticities of output with respect to capital and labor for five industrialized countries—France, West Germany, Japan, the United Kingdom and the United States—without making any restrictive assumptions about the nature of the technology, the behavior of the firms, or the competitiveness of the markets (Boskin and Lau, 1990). We were able to do so by pooling the time-series data of five industrialized countries[4] and making the assumptions that (1) all countries have access to the same technology, that is, they have the same underlying production function, sometimes referred to as a meta-production function;[5] (2) there are differences in the technical efficiencies of production and in the qualities and definitions of measured inputs across countries; and (3) the inputs of the different countries may be converted into "efficiency"-equivalent, or standardized, units of inputs by multiplicative country- and input-specific time-varying augmentation factors.

For example,

$$K^*_{it} = A_{iK}(t)K_{it}; \; L^*_{it} = A_{iL}(t)L_{it}. \tag{3}$$

where K^*_{it} and L^*_{it} are the "efficiency"-equivalent quantities of capital and labor respectively of the ith country at time t; $A_{iK}(t)$ and $A_{iL}(t)$ are the capital- and labor-augmentation factors respectively for the ith country; and K_{it} and L_{it} are the measured quantities of capital and labor of the ith country at time t. The meta- or common production function is valid for all countries in terms of "efficiency"-equivalent quantities of outputs and inputs.[6]

These assumptions require some explanation. Essentially it is assumed that the aggregate production function is the same everywhere in terms of "efficiency"-equivalent or standardized units of inputs. For example, one unit of capital in country A may be equivalent to two units of capital in country *B*; and one unit of labor in country A may be equivalent to one-third of a unit of labor in country *B*. In terms of the measured quantities of inputs, the production functions of the two countries are *not* likely to be the same. However, in terms of "efficiency"-equivalent or standardized units, the assumption of a common production function across countries is far more plausible.

The question inevitably arises: How can one estimate these augmentation or conversion factors? How can one standardize the measured inputs? It turns out that these augmentation factors can in fact be estimated simultaneously with the parameters of the aggregate production function from data on the quantities of measured outputs and inputs, subject to a normalization. Thus, it is actually possible to answer the question of how many units of labor in country *B* is equivalent to 1 unit of labor in country *A* at some given time *t* empirically.

Using this approach, we are able to estimate the rates and patterns of scale economies and technical progress, as well as the relative contributions of the inputs to economic growth, without making the three conventional assumptions, mentioned above, maintained in growth accounting.

TESTING OF HYPOTHESES

An aggregate production function of the transcendental logarithmic form with exponential augmentation factors is specified and estimated using the instrumental variables method.[8] For each country and input, the augmentation factor is assumed to take the form $X^*_{ijt} = A_{ij}\exp(c_{ij}t)X_{ijt}$ where X^*_{ijt} is the (unobserved) "efficiency"-equivalent quantity of the jth input in the ith country at time t, A_{ij} is the augmentation level parameter, c_{ij} is the augmentation rate parameter, and X_{ijt} is the measured quantity of the jth input in the ith country at time t.

A series of hypotheses are tested. The first hypothesis that is tested is that of identical production functions across countries in terms of "efficiency"-equivalent units of inputs. Is the assumption that all of the countries in the sample operate on the same production function a valid one? This hypothesis cannot be rejected at any reasonable level of significance. The second hypothesis that is tested is whether technical progress can be represented in the output and input-augmentation form with exponential augmentation factors. Once again, the hypothesis cannot be rejected. We conclude that our maintained hypothesis of a meta-production function of the commodity-augmenting translog form with exponential commodity-augmentation factors cannot be rejected.

Next, we test the maintained hypotheses of conventional growth accounting: constant returns to scale, neutrality of technical progress and profit maximization. All three hypotheses, as well as the hypothesis of homogeneity of the production function, which is implied by constant returns to scale, can be rejected at the 1 percent level of significance.

Having established the validity of our approach and the lack of validity of the conventional assumptions, we proceed to examine the structure of the technology. We test and cannot reject the hypothesis that the augmentation level parameters for capital and labor are the same across all countries. In other words, in the base period (1970), the "efficiency" levels of capital and labor were not significantly different across countries. We test and cannot reject the hypothesis that technical progress can be adequately represented by two augmentation rate parameters rather than three (output, capital and labor) for each country. We also test and cannot reject the hypothesis that technical progress can be adequately represented by a single augmentation rate parameter for each country. The one single augmentation rate turns out to be that of capital.

We therefore conclude that technical progress can be represented as capital-augmenting, which means, in particular, that technical progress is *biased*. Capital-augmenting technical progress implies that capital and technical progress are complementary: the benefits of technical progress are greater the larger the capital stock, other things being equal.

We also find that the estimated labor elasticities are generally comparable in magnitude to the actual shares of labor cost in total output (even though the hypothesis of profit maximization with respect to labor is actually rejected). However, the estimated capital elasticities are much lower in magnitude than were customarily found or assumed. (Recall that the hypothesis of constant returns to scale is not maintained in our approach.) The low capital elasticities result in estimates of significant decreasing returns to scale in capital and labor within the observed range of the inputs. (Bear in mind that the translog production function is not necessarily homogeneous and therefore does not exhibit fixed returns to scale.) Finally, the implied average annual rates of technical progress, that is, the rate of growth of output holding inputs constant, rank the different countries in the order of Japan (3.9 percent), France (3.1 percent), West Germany (2.7 percent), the United States (1.9 percent), and the United Kingdom (1.8 percent). It is interesting to note that these rates bear a direct relationship to the rates of growth of the capital stock of these countries over the sample period, illustrating the capital-technology complementarity implied by capital-augmenting technical progress (see Figure 1).

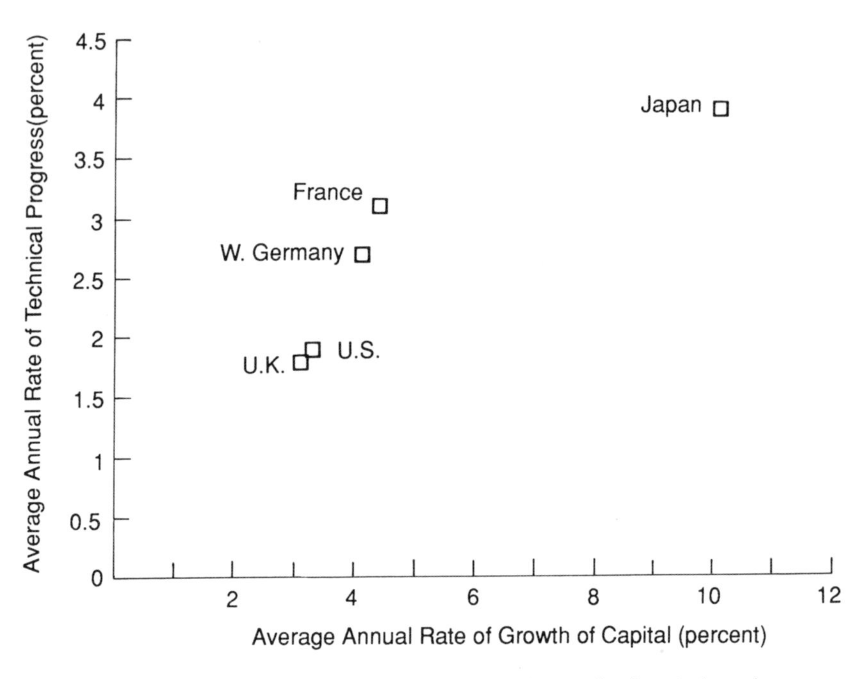

FIGURE 1 Relation between technical progress and growth of capital stock.

ACCOUNTING FOR GROWTH

What do these results imply about the relative contributions of the three sources of economic growth: capital, labor, and technical progress? They indicate that technical progress is by far the most important direct source of economic growth for the industrialized countries in our sample, accounting for more than 50 percent of the growth in real aggregate output (more than 80 percent for the European countries), followed by capital (with the exception of the United States), with labor a distant third. Moreover, capital and technical progress combined account directly for more than 95 percent of the economic growth of the industrialized countries except the United States. In the United States, where labor grew much more rapidly than in the other countries during this period, capital formation and technical progress still account directly for approximately 75 percent of economic growth.

Our results may be contrasted with those of growth accounting exercises using more conventional methods. The conventional approach which assumes constant returns to scale, attributes a much higher proportion of economic growth to capital and a correspondingly lower proportion to technical progress. However, it would be wrong to interpret our finding to mean that capital is not an important source of economic growth. The effect of technical progress is found to depend directly on the size of the capital stock. It is interesting to note that the combined effects of capital and technical progress are virtually the same under both approaches as are the effects due to the growth of labor.

Our identification of technical progress as the most important source of economic growth is reminiscent of the findings of a large unexplained "residual" in early studies of economic growth. Such superficial similarities may, however, be misleading. Our finding is distinct from the earlier ones on at least two counts. First, we do not assume constant returns to scale, neutrality of technical progress and profit maximization with competitive markets. Second, we find empirically, and we believe for the first time, capital and technical progress to be complementary, so that the effect of technical progress on real output depends on the size of the capital stock. The implications of this capital-augmenting type of technical progress are quite different from those of the conventionally assumed neutral variety, the effect of which does not depend on the size of the capital stock.

CONCLUDING REMARKS

We have discussed briefly our method for analyzing important characteristics of the nature of economic growth, such as the rates and pattern of technical progress and scale economies, simultaneously, using pooled time-series data from several industrialized countries. We have found that the empirical data are inconsistent with the hypothesis of constant returns to scale at the aggregate national level. In fact, we have found significant decreasing returns to scale. Moreover, we have found that technical progress is far from neutral. In fact, it is capital-augmenting. In addition, we have also found that the empirical data are inconsistent with the assumption of profit maximization with respect to labor under competitive conditions.

What are the implications of capital-augmenting technical progress?[9] At the aggregate level, one implication of capital-augmenting technical progress is the importance of capital to long-term economic growth. The benefits of technical progress to the economy are directly proportional to the size of the capital stock. A country with a low level of capital stock relative to labor will not benefit as much from technical progress as a country with a high level of capital stock relative to labor. Capital and technical progress are, in a word, complementary. A second implication is that technical progress is more likely to be capital-saving, in the sense that the desired capital-labor ratio for given prices of capital and labor and the quantity of output declines with technical progress, rather than labor-saving. Capital-augmenting technical is thus less likely to cause long-run structural unemployment through the displacement of workers.

We have treated technical progress as exogenous in this study, in the sense that the rates of factor augmentation are assumed to be determined exogenously. The actual technical progress realized, in the sense of the rate of growth of real output, holding inputs constant, is in fact endogenous, as it depends on capital and labor in addition to time. It is, however, remarkable that the rate of augmentation of capital turns out to be almost identical for France, West Germany, and Japan, indicating that the three countries have nearly the same access to advances in technology. It would be of interest to investigate the determinants of the differences in the rates of capital augmentation across countries. Can they be explained by education, R&D expenditures, rate of capital accumulation, or other factors?

The preliminary results presented here are interesting, but much additional work remains. Promising potential areas include accounting for omitted factors such as land (Lau and Yotopoulos, 1989), human capital, public capital (Boskin et al., 1989), R&D and environmental capital (which may explain the significantly decreasing returns to scale to capital and labor observed in the sample); possible vintage effects; and embodiment of technical progress.

ACKNOWLEDGMENT

This brief review reports on some of the results of a major study of postwar economic growth in the five largest industrial economies, conducted at Stanford University in 1987 and 1988. The full, completed study is Boskin and Lau (1990). The authors wish to thank the Ford Foundation, the John M. Olin Foundation, the Pine Tree Charitable Trust, and the Technology and Economic Growth Program of the Center for Economic Policy Research, Stanford University, for financial support. They are grateful to Paul David, Bert Hickman, Lawrence Klein, and Ralph Landau for helpful discussions. The views expressed herein are the authors' and do not necessarily reflect those of the institutions with which they are affiliated.

NOTES

1. See Griliches and Jorgenson (1966), Jorgenson and Griliches (1967), and Jorgenson et al. (1987).
2. In almost all such formulations, technical progress is taken to be exogenous.

3. One exception is Denison (1967), who assumes modestly increasing returns to scale, 1.1 rather than 1 (constant returns to scale).

4. The data are taken from official sources, principally publications of the Organization for Economic Cooperation and Development (OECD) and the U.S. Department of Commerce. See Boskin and Lau (1990) for details. The sample periods are 1957-1985 for France, Japan and the United Kingdom, 1960-1985 for West Germany, and 1948-1985 for the United States.

5. This term is due to Hayami and Ruttan (1970, 1985).

6. This commodity-augmentation modification of the meta-production function was introduced by Lau and Yotopoulos (1989).

7. This functional form was introduced by Christensen, Jorgenson, and Lau (1973).

8. For details, see Boskin and Lau (1990).

9. This is sometimes referred to as Solow-neutral technical change.

REFERENCES

Abramovitz, M. 1956. Resource and output trends in the United States since 1870. American Economic Review 46:5-23.

Boskin, M. J. 1988. Tax policy and economic growth: Lessons from the 1980s. Journal of Economic Perspectives 2:71-97.

Boskin, M. J., and L. J. Lau. 1990. Post-War Economic Growth in the Group-of-Five Countries: A New Analysis. Working Paper. Department of Economics, Stanford University.

Boskin, M. J., M. S. Robinson, and A. M. Huber. 1989. Government saving, capital formation and wealth in the United States, 1947-1985. Pp. 287-356 in R. E. Lipsey and H. S. Tice, The Measurement of Saving, Investment, and Wealth. Chicago: University of Chicago Press.

Christensen, L. R., D. W. Jorgenson, and L. J. Lau. 1973. Transcendental logarithmic production frontiers. Review of Economics and Statistics 55:28-45.

Denison, E. F. 1962a. The sources of economic growth in the United States and the alternatives before us. New York: Committee on Economic Development. Supplementary Paper No. 13. 297 pages.

Denison, E. F. 1962b. United States economic growth. Journal of Business 35:109-121.

Denison, E. F. 1967. Why Growth Rates Differ: Post-War Experience in Nine Western Countries. Washington, D.C.: Brookings Institution.

Gallant, A. R., and D. W. Jorgenson. 1979. Statistical inference for a system of simultaneous, nonlinear, implicit equations in the context of instrumental variables estimation. Journal of Econometrics 113:272-302.

Griliches, Z., and D. W. Jorgenson. 1966. Sources of measured productivity change: Capital input. American Economic Review 56:50-61.

Hayami, Y., and V. W. Ruttan. 1970. Agricultural productivity differences among countries. American Economic Review 60:895-911.

Hayami, Y., and V. W. Ruttan. 1985. Agricultural Development: An International Perspective, revised and expanded ed. Baltimore, Md.: Johns Hopkins University Press.

Jorgenson, D. W., and Z. Griliches. 1967. The explanation of productivity change. Review of Economic Studies 34:249-283.

Jorgenson, D. W., F. M. Gollop, and B. M. Fraumeni. 1987. Productivity and U.S. Economic Growth. Cambridge, Mass.: Harvard University Press.

Kendrick, J. W. 1961. Productivity Trends in the United States. Princeton, N.J.: Princeton University Press.

Kendrick, J. W. 1973. Postwar Productivity Trends in the United States, 1948-1969. New York: Columbia University Press.

Kuznets, S. S. 1965. Economic Growth and Structure. New York: Norton.

Kuznets, S. S. 1966. Modern Economic Growth: Rate, Structure and Spread. New Haven, Conn.: Yale University Press.

Kuznets, S. S. 1971. Economic Growth of Nations. Cambridge, Mass.: Harvard University Press.

Kuznets, S. S. 1973. Population, Capital and Growth. New York: Norton.

Lau, L. J., and P. A. Yotopoulos. 1989. The meta-production function approach to technological change in world agriculture. Journal of Development Economics 31:241-269.

Lau, L. J., and P. A. Yotopoulos. 1990. Intercountry differences in agricultural productivity: An application of the meta-production function. Department of Economics, Stanford University. Photocopy.

Lindbeck, A. 1983. The recent slowdown of productivity growth. Economic Journal 93:13-34.

Solow, R. M. 1956. A contribution to the theory of economic growth. Quarterly Journal of Economics 70:65-94.

Solow, R. M. 1957. Technical change and the aggregate production function. Review of Economics and Statistics 39:312-320.

Michael J. Boskin, an economist and educator, is currently chairman, President's Council of Economic Advisers. He is on leave from Stanford University, where he is the Wohlford Professor of Economics, chairman of the Center for Economic Policy at Stanford University, and research associate, National Bureau of Economic Research. He is the author of 2 books and approximately 50 articles and editor of 6 volumes of essays on taxation, fiscal policy, capital formation, labor markets, social security, and related subjects. The recipient of numerous honors and awards, Dr. Boskin received his B.A., M.A., and Ph.D. degrees from the University of California, Berkeley.

Lawrence J. Lau is professor of economics at Stanford University. Since joining the Stanford faculty in 1966, he has been a visiting assistant research economist at the University of California, Berkeley, and a visiting professor of economics at Harvard University. He graduated from Stanford University in 1964 with a Bachelor of Science degree with Great Distinction in physics and economics. He later received both master's and doctorate degrees in economics at the University of California, Berkeley. He is the author of several books and approximately 100 articles in economics.

Investing in Productivity Growth

Dale W. Jorgenson

More rapid growth in productivity is essential for achieving the goals of U.S. economic policy. The slowing of U.S. economic growth in the 1970s can be attributed in large part to the decline in productivity growth. Productivity growth is an important component of the increase in our standard of living. More recently, the rise of the U.S. current account deficit in the 1980s is often ascribed to more rapid productivity growth in other countries, especially Japan and the Four Dragons of East Asia—Hong Kong, Korea, Singapore, and Taiwan. Productivity growth is necessary for enhancing our international economic competitiveness.

It is important to recognize at the outset of our discussion that the productivity problem is enormously complex, involving the performance of our whole economy and, in some ways, our whole society. The solution of the productivity problem will require rethinking our approach to economic policy. A new approach to economic policy is suggested by the idea that improving the performance of the U.S. economy requires investing in productivity. Newly available data on productivity make it possible to identify opportunities for productivity-enhancing investments. These investments will take many forms, but we can identify three that are critical to future productivity growth:

1. Tangible assets. This is the conventional meaning of investment. It includes investment in plant and equipment for the business sector, housing and consumers' durables for the household sector, and military equipment and civilian infrastructure for the government sector. It is useful to think of this as investment in *hardware*.
2. Intangible assets. Investments require the commitment of capital

resources and produce changes in technology that promote productivity growth. However, not all investments take the form of bricks and mortar. A rapidly growing portion of investment takes the form of research and development, advertising and marketing, and intangibles such as computer software. I will refer to investment in intangibles as *software*.

3. Human capital. The most important component of investment in human capital is schooling. Formal schooling extends all the way from kindergarten to the most specialized forms of higher education. This was the focus of the Charlottesville education summit organized by President Bush in the fall of 1989. However, investment in human capital also includes training on the job. Mincer (1989) has estimated that training costs in the United States amount to 35-42 percent of schooling costs. I find it useful to refer to education and training as investments in *people*.

My next objective is to develop a perspective on opportunities for investing in productivity. For this purpose I will summarize data on productivity that are presented in more detail in my book with Frank Gollop and Barbara Fraumeni, *Productivity and U.S. Economic Growth*, published by the Harvard University Press in 1987. A very important feature of these data is that we can identify specific channels for the impact of investments in hardware, software, and people on productivity growth.

Productivity is the ratio of output to input. I use gross domestic product (GDP) as a measure of output. This includes the output of all economic activities in the United States, whether conducted by Americans or foreigners. By contrast the gross national product (GNP) is output by Americans, whether at home or abroad. I will take hours worked as a measure of input, since this represents the most rudimentary measure of effort. Productivity is defined as gross domestic product per hour worked. This output is produced by combining labor in the form of hours worked with all the forms of capital I have mentioned—hardware, software, and people.

At this point I need a distinction that is crucial in interpreting productivity growth. This is between the productivity growth that can be attributed to investment and productivity growth that does not require investment. Economists refer to the first as "explained" by the commitment of greater resources and the second as the "unexplained" residual. It is important to understand the basis for this distinction, since it is central to deriving the implications of productivity growth for economic policy.

We can illustrate the distinction between explained and unexplained productivity growth by the familiar example of mechanization. If we replace 10 people with 10 shovels by 1 person with a power-driven digging machine, we have mechanized the digging process. Economist's jargon for this transformation is to say that capital in the form of hardware is substituted for labor in the form of hours worked. This is a simple but useful illustration of productivity growth as investment in hardware. Output per hour worked increases as a result of investment in construction machinery.

A summary of U.S. productivity growth over the period 1947-1985 is presented in the following table:

Productivity Growth (percent)

U.S. Productivity Growth, 1947-1985	1947-1985	1979-1985
Growth Rate (GDP/hours worked)	2.10	1.05
Labor Quality	0.39 (19%)	0.29 (28%)
Capital Quality	0.58 (28%)	0.31 (30%)
"Capital-Labor" Substitution	0.41 (19%)	0.14 (13%)
R&D (High)	0.25 (12%)	0.25 (24%)
Residual	0.46 (22%)	0.06 (6%)

The explanation of productivity growth as investment in hardware, software, and people is the key to deriving the implications for economic policy. In the example of mechanization, productivity growth results from investment in hardware. We need additional saving and investment to obtain gains in productivity. However, not all investments involve hardware and not all productivity growth involves investment. The data in the table provide the latest information on sources of growth in productivity. To interpret the results I need to relate investments in hardware, software, and people to the example of substitution of capital for labor.

First, the simplest form of substitution of capital for labor is the substitution of hardware for hours worked. In the table this is labeled "capital-labor" substitution. The quotation marks indicate that this is only one of many possible forms of substitution of capital for labor. This accounts for about one-fifth of the productivity growth (19 percent to be precise) during the postwar period. We can also substitute people for people to increase productivity. This results from hiring more highly educated and trained people in place of less well educated and trained people. Education and training both require substantial commitments of resources to produce growth in productivity. Investments in people, which are labeled growth in labor quality in the table, account for about one-fifth of productivity growth (19 percent) over the postwar period.

The third kind of substitution included in the table is substitution among different types of hardware or tangible capital. Obviously, machines differ in effectiveness, just as people do. By increasing our investment in hardware we can obtain more effective plant and equipment, which is a very important form of substitution. This is encompassed in the measure of capital quality given in the table. Growth in capital quality accounts for more than a quarter (28 percent) of growth in productivity during the postwar period. This is almost 50 percent more important than growth in labor quality or "capital-labor" substitution. Growth in capital quality and "capital-labor" substitution constitute the contribution of investment in tangible assets to productivity growth. This amounts to almost half (47 percent) of productivity growth and is by far the most important source of increases in productivity.

The contribution of investments in hardware and people together make up nearly two-thirds (66 percent) of growth in productivity during the postwar period. To complete the picture, we require an estimate of the contribution of investment in software to productivity growth. For this purpose we use a "high" estimate of the contribution of research and development presented by Griliches (1988). This estimate is based on a detailed analysis of data on research and development investment by individual firms. The contribution of investment in software to productivity growth (12 percent) is considerably less than that of investment in people (19 percent). It is important to keep in mind that this can be regarded as an upper bound to the contribution of research and development expenditures.

Up to this point we have focused attention on the sources of growth in productivity that can be attributed to the commitment of additional resources through investments in hardware, software, and people. This leaves an "unexplained" residual, which accounts for about one-fifth (22 percent) of productivity growth. Abramovitz (1962) has referred to this residual as the Measure of our Ignorance. Our overall conclusion is that almost four-fifths (78 percent) of productivity growth can be explained by investments in hardware, software, and people. Growth in productivity is primarily the result of mobilizing investment resources and deploying them efficiently. The presence of the unexplained residual is useful in reminding us that additional economic research on the sources of productivity growth will be needed to provide a complete explanation.

The first line of this table shows that productivity growth has averaged 2.10 percent per year over the postwar period 1947-1985. This estimate is based on my estimates with Gollop and Fraumeni (1987), brought up to date in my recent paper (1991). Since 1979, the growth rate of productivity has been halved, running at only 1.05 percent.

This is about half the average rate of productivity growth during the postwar period, 1947-1985. An important part of the explanation of this decline is the near disappearance of the "unexplained" residual. This contribution to productivity growth has fallen from 0.46 to 0.06 percent per year and accounts for only 6 percent of productivity growth during the period 1979-1985.

For this period the relative importance of investments in hardware, software, and people in explaining productivity growth is much greater than for the entire postwar period. However, the role of these three components of investment has changed. For example, the contribution of "capital-labor" substitution, trading off additional capital against fewer hours, has fallen dramatically from 0.41 percent to 0.14 percent per year. Growth in capital quality has only declined from 0.58 to 0.31 percent per year. Taking these two contributions together, the relative importance of investment in tangible assets has declined from 47 to 43 percent of productivity growth.

Unfortunately, we do not have the data required to estimate the contribution of investment in research and development to productivity growth separately for the period 1979-1985 and the postwar period as a whole. We assume that this contribution remains unchanged, which is consistent with the findings reported by Griliches (1988). By contrast the contribution of investment in human capital has declined from 0.39 to 0.29 percent per year, but the relative importance of this investment has risen from 19 to 28 percent of productivity growth. Our assumption that the contribution of investment in research and development has remained unchanged produces a doubling of the relative importance of this form of investment. This is the consequence of the halving of productivity growth during the period 1979-1985.

We now come to the bottom line: What is the contribution that economic policy can make to productivity growth? The answers are:

1. Investment in hardware. The government can design a tax system that will promote saving and investment and ensure that capital is allocated efficiently among alternative uses. The government can also design economic regulation in such a way as to minimize loss in efficiency. The government also has special responsibilities for investment in civilian infrastructure and defense equipment. The role of these investments in productivity growth is not well understood, at least by economists, and is allocated to the unexplained residual in the estimates we have presented.

2. Investment in software. The federal government can promote investments in research and development, advertising and market-

ing, and other forms of intangible assets. The government finances an important part of research and development and conducts research in its own laboratories in such areas as health, energy, and defense. Efficient allocation of the funds available for these forms of investment is an important issue in science and technology policy.

3. Investment in people. The most important investments in people are made in the educational system. State and local governments have the main responsibility for these investments. However, the largest part of the cost of these investments is contributed by individuals through the time and effort that they devote to education.

Economists have not arrived at a complete understanding of the generation of productivity growth. However, the picture that has emerged from recent research is clear enough. Gains in productivity are due primarily to investments in hardware, software, and people. To stimulate productivity growth we need to mobilize capital through savings and deploy the resulting investments as efficiently as possible. Decisions to save and invest take place in business, household, and government sectors. Businesses accumulate tangible and intangible assets and invest in human capital through on-the-job training. Households make important investments in housing. They also invest in human capital through undertaking formal education and undergoing on-the-job training.

Ensuring continued growth in productivity involves a wide range of government policies: tax and regulatory policies for tangible assets, science and technology policies for intangibles, and education and training policies for human capital. The first part of the solution to the productivity problem is to generate new investments in hardware, software, and people. Equally important is to allocate these investments in the most efficient way. Only by using our scarce investment resources efficiently can we obtain the productivity gains that are essential to growth in our standard of living and restoration of our international economic competitiveness.

REFERENCES

Abramovitz, M. 1962. Economic Growth in the United States. American Economic Review 52(4)(September):762-782.

Griliches, Z. 1988. Productivity puzzles and R&D: Another nonexplanation. Journal of Economic Perspectives 2(4)(Fall):9-22.

Jorgenson, D. W. 1991. Productivity and economic growth. In Measurement in Economics, E. Berndt and J. Triplett, eds. Chicago: University of Chicago Press.

Jorgenson D. W., F. M. Gollop, and B. M. Fraumeni. 1987. Productivity and U.S. Economic Growth Cambridge, Mass.: Harvard University Press.

Mincer, J. 1989. Job Training: Costs, Returns, and Wage Profiles. Department of Economics, Columbia University, September.

Dale W. Jorgenson is Frederic Eaton Abbe Professor of Economics, Harvard University. He received his educational training from Reed College (B.A.) and Harvard University (A.M., and Ph.D.). A few of his academic appointments have included director, Program on Technology and Economic Policy, Kennedy School of Government, Harvard University; professor of economics, University of California, Berkeley; visiting professor of economics, Stanford University; visiting professor of statistics, Oxford University; and Ford Foundation Research Professor of Economics, University of Chicago. Dr. Jorgenson has authored and collaborated on approximately 180 publications on the topic of economics. He is a member of the National Academy of Sciences and a foreign member of the Royal Swedish Academy of Sciences.

Technology and the Cost of Equity Capital

George N. Hatsopoulos

In 1985, when the U.S. trade deficit reached an unprecedented 3 percent of U.S. gross national product and Japanese trade surpluses reached an equally unprecedented 4 percent of Japan's gross national product, economists argued that the principal cause of the problem was the abnormally high exchange rate of the dollar at 260 yen. Many economists contended that our trade balance would be restored simply by reducing the exchange rate to about 170 yen per dollar. At about the same time, our business leaders were addressing the trade issue in their own way. In their view, U.S. manufacturers could not be competitive when Japanese wages were only about half those prevailing here.

Five years later, with the value of the dollar now reduced by almost a factor of two and with Japanese wages having reached parity with our own, the U.S. trade deficit persists. The level is somewhat reduced, but, most significantly, Japanese surpluses continue unabated. Today few people doubt that U.S. industrial competitiveness is declining; there is, however, as yet, no consensus on the causes of such decline. Nevertheless, a central issue seems to attract growing attention: the shortsightedness of U.S. corporate management. This issue, which has been discussed in academic circles for some time, attained national focus in early 1987 as the result of a speech by Richard Darman, then deputy secretary of the treasury. Further support for this position was given by the Massachusetts Institute of Technology report on industrial competitiveness, *Made in America* (Dertouzos et al., 1989).

The MIT study presents ample evidence of American management's preoccupation with short-term profits and its lack of commitment to long-term competitiveness. This is not news to most Americans.

Consumers long have been aware of declining quality of U.S.-made products compared with imports. Moreover, there is growing awareness that foreign competitors, particularly the Japanese, are more willing to forgo current profits to increase market share in virtually any manufacturing enterprise. National data on industrial investment in research and development corroborate these perceptions. From 1986 to 1989, spending for research and development by U.S. corporations declined as a share of gross national product. In fact, total research and development spending, adjusted for inflation, actually declined in 1989.

We see a lot of technology being developed in this country, ranging from basic technology to practical inventions. Yet, America incorporates its innovation and technology into internationally marketed products and services at a rate that is scandalously below its true potential. Japan appropriates and commercializes American technology at a much higher rate than we do. The thesis presented in this paper is that the overriding factor affecting the ability of an industry to incorporate technology into products is the rate of return on equity capital demanded by stockholders.

The average price-to-earnings ratio for corporations listed on the New York Stock Exchange is currently about 14. The corresponding figure for the Tokyo Stock Exchange, corrected for accounting differences, is roughly 40. This means that a U.S. corporate manager must ensure that for every $100 of equity investment, the annual return will be at least $7. In Japan, the corresponding figure is $2.50. It also means that any equity-financed investment that promises to double in value within 12 years would be considered an irresponsible act in the United States, but worthy of praise in Japan. Small wonder that Japanese managers take a longer view!

The differing economic environments that drive both investment decisions and planning horizons in various countries can be described by a variable known as the *cost of capital*. In the following discussion, we shall examine what cost of capital means, what it affects, and what it is affected by. Recent estimates of the cost of capital in the United States and Japan also are presented.

THE COST OF CAPITAL

In an ideal world, free of both taxes and risks, corporations could finance any and all projects through borrowing, provided that those projects were sure to return more than the interest rate demanded by lenders. In such a world, the cost of capital would be the interest rate. Moreover, if competition were "perfect," the return on all projects would be the same, and it would be identical to the cost of capital.

In the real world, of course, things are quite different. Taxes have to be paid under rather complicated rules, and predictions cannot be made with certainty. Therefore, all projects involve risks, and those risks represent a significant burden to prospective investors. At some price, lenders are willing to bear a certain level of risk. In practice, most of that risk is borne by holders of equity.

In the imperfect world, equity holders are represented by corporate managers whose fiduciary obligation is to provide, as best as they can, the return required by these holders. It follows that corporate managers should invest only in those projects that promise to return a pretax profit sufficient to provide the taxes required by law, the interest required by lenders, and dividends and capital gains required by equity holders.

The real net cost of capital, or simply the cost of capital, for an investment is the least return that satisfies all of the requirements cited above. Its magnitude depends on many factors, including tax rates. As a result, the cost of capital for a given type of investment by a corporation may differ from that of an individual, simply because corporations are taxed at different rates than individuals. The cost of capital that is important to the international competitiveness of nations, however, is that which pertains to corporations, because most international trade is done by corporations.

The cost of capital provides a criterion for investments in new projects. Actual returns on past investments may differ from the current cost of capital as a result of changes in the economic environs. It can even differ from that prevailing at the time those investments were made, since unforeseen events may have raised or lowered the return on past projects.

A general expression for corporate cost of capital is given in Hatsopoulos and Brooks (1985). Its derivation is based on the approach developed by Hall and Jorgenson (1967). The expression contains several variables, principally the real after-tax cost of funds described below, the depreciation rate of the investment, and several elements of the tax code.

In making investments, corporations use two sources of funds—equity and debt. Each source differs in its exposure to risk, its taxation, and its cost.

The use of equity exposes a corporation to the least risk, because it involves no fixed obligation to provide either returns or repayments. For the same reason, the supplier of equity funds is exposed to the biggest risk. Use of interest-bearing debt exposes a corporation to the biggest risk, and the supplier of funds to the least, because it involves a fixed obligation to provide returns and to repay the funds. The two sources of funds impose different corporate tax burdens on the return to holders. Payments to equity holders are taxed, whereas payments to debt holders are not.

The real after-tax cost of funds is the average of the real after-tax cost of debt and the real cost of equity. These are weighted in proportion to the relative amounts of debt and equity used by the corporation to finance a given investment.

The real after-tax cost of debt, Cd, is related to the interest rate, i, which the corporation must pay to lenders. This rate is adjusted for inflation and for the tax deduction on interest payments provided by the tax code. The adjustments yield the following expressions for Cd:

$$Cd = i\,(1 - \tau) - \pi$$

where τ is the tax rate on corporate income and π the inflation rate.

The real after-tax cost of debt is always much smaller than the interest rate. In fact, it occasionally becomes negative. Table 1 lists the average interest rate and the real cost of debt for U.S. nonfinancial corporations in selected years.

TABLE 1 The Cost of Corporate Debt and Equity in the United States, percent

	1974	1981	1988
Interest rate on AAA bonds, i	8.60	14.20	9.70
Inflation rate[1], π	8.60	9.20	3.20
Real interest rate[2], i^*	0	5.00	6.50
Corporate tax rate[3], τ	0.52	0.50	0.42
Real cost of debt after taxes[4]	–4.50	–2.10	2.40
Nominal cost of equity	16.20	16.30	9.50
Real cost of equity	7.60	7.10	6.20

NOTES:
(1) Rate of change of the GNP deflator, fourth quarter to fourth quarter
(2) $i^* = i - \pi$
(3) Including federal and state taxes.
(4) C.O.D. = $i(1 - \tau) - \pi$

THE COST OF EQUITY

Stockholders invest in corporate equities in order to have future monetary returns. Corporations invest the stockholders' equity to make an after-tax profit. Part of that profit is paid back to investors, and part of it is retained for reinvestment to generate progressively larger profits that will allow progressively higher cash payments to stockholders. A corporation's cost of equity is the rate of return stockholders demand.

The concept is analogous to the definition of the interest rate, that is, the rate of return that lenders demand. There is, however, an important difference. Whereas the cost of debt to a corporation is less than the interest rate, the cost of equity in America is the same as the return demanded by stockholders. This is because no corporate tax deduction is provided for payments made by the corporation to its stockholders.

The phrase "the return stockholders demand" used in the above definition, although widely used in economics, may sound strange to anyone but an economist. Everyone knows how lenders can enforce their demand to receive interest payments from a corporation. A loan constitutes a contractual obligation between the lender and the borrower, and if the borrower does not pay the contracted return, the lender can take legal action.

On the other hand, equity holders have no legal recourse when a company fails to earn what stockholders require. They can, however, bid down the shares of the company until the share price reflects the earning power of the company, and hence, the required rate of return. A continuing market valuation of a company much below the replacement value of its net assets will at best deprive the company of access to new equity capital. At worst, the management may be replaced. This can occur either through the action of the company's board or, as often happens today, through an unfriendly takeover. To prevent such actions, managers will tend to increase payout in the form of dividends or stock repurchases and will reduce new investments.

Empirical determination of the cost of equity is a somewhat complicated and frequently misunderstood process. To illustrate the point, consider the following example: Assume a company has net income of $50,000 per year, and the replacement value of its assets, less its liabilities, is $1 million. Thus, its return on equity is 5 percent. Let us further assume that the company pays out to its stockholders $20,000 per year, reinvests $30,000 per year at its historical return of 5 percent, and has a cost-of-equity equal to 10 percent. Under these assumptions, the market value of the company's stock will be $350,000. This reflects the $20,000 of distributed earnings that is valued at 10 to 1, plus $150,000 that represents $30,000 reinvested at half the required return. Thus, the market value of the company's equity is only 35 percent of its replacement value, or 70 percent of the value of its current earnings divided by the required return. The return on the company's market value is $50,000/$350,000, or 14.3 percent and, the price-to-earnings ratio of the company is 7, namely, the inverse of 0.143.

It is evident from the above that the cost of equity (10 percent) differs from the return on equity at replacement (5 percent), and from the inverse of the price-to-earnings ratio (14.3 percent). In general, these three parameters will also differ for any real-world company.

CORPORATE HORIZONS AND THE COST OF CAPITAL

The cost of capital divided by the cost of labor is a principal determinant of the capital-to-labor ratio in an industry. The capital-to-labor ratio, in turn, is an important determinant of labor productivity, and this is a key determinant of standard of living. These propositions are well accepted by economists, although there is disagreement concerning the extent to which capital-to-labor ratios affect productivity. Historically, economists define "capital" as hard assets such as plant and equipment, not soft assets such as technology. In recent years, however, such soft assets have become increasingly important. In fact, they may well be the most important type of capital for the industrial competitiveness of a high-wage country such as the United States.

The reason for this shift in emphasis is that several countries, West Germany and Japan in particular, have reached a stage of technological development sufficient to compete with the United States in industries that are technology-intensive. Such industries command a higher price per unit of labor than do commodities-oriented industries, because fewer countries can compete. In addition, there are markets such as consumer electronics in which market dominance and product quality can provide a competitive edge sufficient to allow market leaders to command higher prices. Achievement of such dominance and quality also requires heavy intangible investments in marketing and product quality. Thus, for a high-wage country to be competitive, the most important attribute is to have a low cost of capital for soft or intangible investments.

The problem of America's competitiveness in international trade is largely related to manufacturing. Changes in the trade balance in manufacturing were the principal cause of the U.S. trade deficit in the 1980s. Moreover, improved trade in products and services with high technological content is likely to be the principal means of reducing future U.S. trade deficits. For such products and services, intangible investments are more critical to international competitiveness than are investments in fixed assets.

As noted earlier, the cost of capital in general depends on the cost of debt, the cost of equity (which is much higher than that of debt), and the tax code parameters. Tax variables include the tax rate on corporate income, the investment tax credit, and the depreciation rate allowed for a given type of asset. For soft investments, however, the cost of capital is simply the cost of equity. This fact has important implications regarding national economic policy.

There are two separate reasons why the cost of capital for soft investments is identical to the cost of equity. The first is that the tax codes of all countries provide for immediate expending of such investments. Because of that fact, the cost of capital becomes the after-tax cost of funds. In addition, because it is nearly impossible for lenders to assess the value of soft assets with any degree of certainty, such assets do not provide a suitable collateral for debt financing.

The second reason is based on the fact that accounting rules in all countries require that soft investments be charged against current income. A reduction of current income can be justified to stockholders only if they perceive that such an investment is likely to produce sufficient income in the future that its discounted present value is greater than the cost of the investment. The applicable rate for such a discounting exercise is the cost of equity.

We see, then, that the after-tax cost of equity controls not only the amount of the soft investments that a corporation makes, but also the time horizons of such investments.

On the one hand, for a company or an industry to attain a significant advantage over its competitors, long-term investments are required. For example, it took more than 15 years for Japanese automakers to penetrate the U.S. market. Similarly, it may take more than 15 years to create a market for high-definition television. On the other hand, the longer it takes for a soft investment to produce profits, the greater will be the burden of a high discount rate. Thus, an investment decision that lowers earnings by $1 now, yet would raise earnings by $2, 15 years from now, is a profitable decision only if the firm's cost of equity is 5 percent or less. The same investment will lower the firm's value if its cost of equity is more than 5 percent.

For a high-wage country, time horizons may very well be more important than the rate of investment in determining the nation's competitiveness. If this is true, then the cost of equity in and of itself is more important to the future of our nation than factors such as corporate tax rates, which affect the cost of hard assets.

INTERNATIONAL COMPARISONS

Comparisons of the cost of capital in the major industrialized nations have been made by several investigators. Comparisons of the United States and Japan have attracted particular attention because Japan is viewed as the principal challenger to America's post-World War II industrial supremacy.

Most of the studies that have appeared in the literature[1] have a major weakness: They consider only the effects of interest rates and taxation. They omit other effects, such as risk, that have a dramatic effect on the cost of equity. Of the empirical studies that address the cost of equity in the United States and Japan, the most recent are Hatsopoulos and Brooks (1987), Ando and Auerbach (1989), McCauley and Zimmer (1989), and Bernheim and Shoven (1989). All of these studies find the U.S. cost of capital to be higher than that of Japan.

What emerges from these studies is the following: Nominal interest rates are higher in the United States than in Japan, for example, 9.7 versus 4.3 percent in 1988.[2] After adjusting for inflation, however, the difference has been much smaller in recent years. Corporate taxation is stiffer in Japan than in the United States, which means that if all other factors were the same, the cost of capital should have been higher in Japan. All other factors, however, are not the same.

The first study to include risk, Hatsopoulos (1983), found that the principal factor driving the U.S.-Japan cost-of-capital gap was the high leverage of Japanese corporations. The last year considered in that study was 1981. Since then, Japanese leverage has declined, and U.S. leverage has increased.[3] This convergence, as well as the previously mentioned convergence in real interest rates, has led some observers to conclude that the cost of capital in the two countries must also have converged. It has not.

In the late 1970s, interest rates were lower in Japan than in the United States. But, because of differences in inflation rates, the real after-tax cost of debt was lower in the United States. The Japanese cost-of-capital advantage derived primarily from their much higher leverage, which meant that greater weight in the average cost of capital was placed on low-cost debt as opposed to high-cost equity. During the 1980s, the decline in the Japanese leverage advantage was offset by a decline in the U.S. advantage relative to real after-tax cost of debt. (The real after-tax cost of debt increased in both countries, from negative values to roughly equivalent levels of 2 percent after 1984.) Over the same period, the real after-tax cost of equity remained constant at about 2.5 percent in Japan and about 7 percent in the

United States. As a result, the gap in the cost of capital between the two countries was the same in 1988 as it was in 1979.

Japan's exceptionally low cost of equity is truly startling. It is lower than interest rates and is now comparable to the real after-tax cost of debt. This may very well be the primary reason that the leverage of Japanese corporations has declined—it is now more advantageous to raise equity than to borrow money in Japan. The causes for such a low cost of equity have preoccupied many financial experts. Some have speculated that the high price of corporate shares in Japan is the result of escalation in land prices. This explanation is not convincing, for the observed decline in the discount rate of corporate equities is bound to be reflected in the rise in both the price of shares and the price of scarce corporate land.

The rate of personal taxation on capital gains from corporate shares, together with certain features of Japan's financial structure, are more rational explanations. Until mid-year 1989, the tax rate on capital gains from equities was zero,[4] whereas the maximum tax rate on all other capital gains, such as those resulting from the sale of real estate, was 50 percent. Thus, the enormous Japanese personal saving rate, typically 15 to 20 percent of disposable income, is directed in its entirety by the tax code toward corporate equities. In addition, the financial structure in Japan provides for a broad sharing of corporate risk throughout the society. More than half of the stock of an average Japanese corporation is held by other corporations and by banks. Banks, in turn, have been consistently supported by the Bank of Japan.

The virtual elimination of barriers to capital flows from nation to nation has led many analysts and economists to believe that differences in cost of capital soon will be eliminated. Indeed, as we saw earlier, real interest rates in the United States and Japan have been converging. The same is true to some extent for other countries. Yet, the cost of equity in different countries is not converging. It appears that although funds to finance government bonds or bonds issued by major corporations tend to flow rather freely across borders, funds to finance equities, especially equities other than those of a few major corporations, do not flow freely. During the 21 months from January 1988 through September 1989, foreign investors acquired a net $336 billion of U.S. financial assets but only $4 billion of U.S. traded equities. Such disparity is understandable; except for a few large multinational corporations, most foreigners are not familiar with publicly traded companies in the United States.

In addition to this lack of familiarity and trust on the part of foreign equity investors, there is the problem that corporate risks differ from country to country because of differences in financial structure.

It is apparent, for reasons previously noted, that corporations in Japan embody less risk for their stockholders than do equivalent corporations in the United States.

The dramatic difference in the cost of equity between the United States and Japan raises serious concerns about our ability to maintain a high standard of living for our workers. If our cost of equity—that is, the rate at which U.S. managers must discount the future—does not become competitive with that of the major industrialized nations, the United States will have to surrender those industries that critically depend on long-term investments ranging from worker training to technology. Such a surrender will force us to compete with a growing number of low-wage countries that produce only commodities and mundane products and services.

POLICIES THAT REDUCE THE COST OF EQUITY

The principal factors that affect the cost of equity are the rate of domestic saving, the structure of personal taxes, and the structure of the financial markets.

Increased domestic saving reduces the cost of equity, not only because it reduces interest rates and makes equity investments more attractive, but also because domestic saving rather than foreign saving is the principal source of equity funds. This means that a reduction of the federal deficit, the major detractor from domestic saving, can have significant beneficial effects in the long term, provided such a reduction does not adversely affect private saving.

The structure of personal taxation also can have an important effect on the cost of equity. For example, if the double taxation of retained earnings and dividends is reduced, it is almost a certainty that the cost of equity will fall. The double taxation of retained earnings can be reduced through a tax exclusion for part of any future gains on equity investments. The double taxation of dividends can be reduced through a partial deductibility of dividends from corporate taxable income. Japan has successfully practiced both of these methods.

An alternative approach is the integration of personal and corporate taxation as, for example, is already being done for Subchapter S corporations. Either of these methods would reduce revenues to the government. There are ways, however, to offset such revenue losses through other changes in the tax code. An increased corporate tax rate, for example, would have a relatively small detrimental effect on fixed capital formation, but would preserve all the benefits for intangible capital formation.

CONCLUSIONS

The most important attribute required to sustain growth in America's standard of living in a competitive world is for our corporations to match the investment horizons of foreign competitors. To accomplish this, it is necessary—although probably not sufficient—to reduce the rate at which U.S. corporations discount future earnings. In fact, that discount rate must be made comparable to rates prevailing in the most competitive industrialized nations. Such convergence in discount rates will not occur automatically through the opening up of capital markets. Convergence requires appropriate economic and tax policies that gradually eliminate the current bias favoring consumption and debt relative to equity finance.

NOTES

1. For example, King and Fullerton (1984) and Bernheim and Shoven (1987).
2. The rates cited are for Moody's AAA bonds in the United States and for Nippon Telephone and Telegraph bonds in Japan, respectively.
3. See McCauley and Zimmer (1989).
4. The 1989 tax reform in Japan provides for a taxation on capital gains from equities which ranges from 2 percent for an asset that has doubled in value, to 20 percent for an asset that has had a small increase in value.

REFERENCES

Ando, A., and A. J. Auerbach. 1990. The cost of capital in Japan: Recent evidence and further results. Paper presented at the conference on "Corporate Finance and Related Issues: Comparative Perspectives, Tokyo, Japan, 7-8 January 1990.

Bernheim, B. D., and J. B. Shoven. 1986. Taxation and the cost of capital: An international comparison. Paper presented at the American Council for Capital Formation Conference on Consumption Tax: A Better Alternative? 3 September 1986.

Dertouzos, M. L., R. K. Lester, and R. M. Solow. 1989. Made in America: Regaining the Productive Edge. Cambridge, Mass.: MIT Press.

Hall, R. E., and D. W. Jorgenson. 1967. Tax policy and investment behavior. American Economic Review 57(3)(June): 391-414.

Hatsopoulos, G. N. 1983. High cost of capital: Handicap of American industry. Paper presented at American Business Conference and Thermo Electron Corporation, 26 April 1983.

Hatsopoulos, G. N., and S. H. Brooks. 1986. The gap in the cost of capital: Causes, effects, and remedies. Pp. 221-280 in Technology and Economic Policy, R. Landau and D. Jorgenson, Ballinger eds. Cambridge, Mass.: Ballinger.

Hatsopoulos, G. N., and S. H. Brooks, 1987. The cost of capital in the United States and Japan. Paper presented at the International Conference on the Cost of Capital, Kennedy School of Government, Harvard University, 19-21 November 1987.

McCauley, R., and S. Zimmer. 1989. Explanations for International Differences in the Cost of Capital. New York: Federal Reserve Bank of New York.

GEORGE N. HATSOPOULOS is the founder, chairman of the board, and president of Thermo Electron Corporation, a company that manufactures instruments, cogeneration systems, process equipment, and biomedical products and provides environmental and metallurgical services. He received his education at the National Technical University of Athens and the Massachusetts Institute of Technology, where he received his bachelor's, master's, engineer's, and doctoral degrees in mechanical engineering. He served on the faculty of MIT and has continued his association with the Institute, currently serving as senior lecturer. He is principal author of *Principles of General Thermodynamics* and *Thermionic Energy Conversion*, Volumes I and II. He has published more than 60 articles in professional journals on thermodynamics, energy conversion, energy conservation, energy productivity, capital formation, cost of capital, and the international competitiveness of American industry. Dr. Hatsopoulos is a member of the governing council of the National Academy of Engineering and is a fellow of the American Academy of Arts and Sciences, the American Society of Mechanical Engineers, and Institute of Electrical and Electronics Engineers.

Science and Its Applications: How to Succeed

Michel Boudart

The research enterprise is defined as the organization that generates science and its applications. Success in this endeavor depends on a combination of attitudes that are rarely found in a single individual but may be assembled among members of a winning team. Some of these attitudes are examined in this essay based on external advice and personal reflection.

THE RESEARCH ENTERPRISE

Many arguments over semantics could be avoided if the following words of Louis Pasteur were more widely disseminated:

> There are not two sciences. There is only one science, and the application of science, and these two activities are linked as the fruit is to the tree.

To separate science from its applications is most often counterproductive. Many leading scientists were deeply involved in applications: Louis Pasteur, Lord Kelvin, Walther Nernst. Some famous industrialists were fascinated by science, or perhaps started as scientists: Ernest Solvay, Charles L. Reese, the Varian brothers. Ralph Landau started from chemistry and its applications, developed and commercialized several innovative processes, and is currently devoting much of his energy to the science of economics as related to technology.

Thus, there exists a continuous spectrum of related activities from long-range, fundamental, basic, pioneering, academic, and corporate research to short-range, applied, mission-oriented, industrial research. Whether the motivation of the work is the *right* to know rather than the *need* to know, curiosity rather than the marketplace, good science can be recognized whether it is far removed from or very close to its applications. Good science never drills a dry well: It leads to discovery, invention, or innovation. Its product is a scientific paper, a patent, or high technology. I define good science as successful science wherever it is managed, directed, or conducted from the viewpoint of the student, scientist, or engineer doing it. Unfortunately, there is also bad science that does not add to the pool of knowledge and does not lead to applications. My remarks shall be confined to small science, done in a group of 15 people or so, in opposition to big science. The examples will be borrowed from chemistry and its applications because that is the field I know firsthand.

The question I shall try to answer is one that is of current interest: how to attract the best creative talent to science and its applications? One answer is, by managing success in the research enterprise, be it a research group at the university, a research institute, a national laboratory, or an industrial company.

Yet, there is a fundamental difference between a corporate entity and academe. Milton Friedman stated that the social responsibility of business is to maximize profit within the rules of the game. Perhaps one could add that the social responsibility of academe is to formulate and disseminate the rules of the game. Thus, there are differences in the style of management of a research team, depending on its social responsibility. In looking for proven reasons behind success, I have restricted myself to one field—catalysis, and catalytic technology. I have borrowed from six industrial leaders in the field six key ideas that, in part, contributed to their success. My sources, in alphabetical order, are: Heinz Heinemann, Jim Idol, Wolfgang Sachtler, John Sinfelt, Haldor Topsoe, and Paul Weisz. I have distilled the elements of their success from talks they gave at Stanford University in 1983 on their philosophy of research. These six ways to success in catalytic science can be expressed by six short exhortations: Exploit luck through observation, optimize chaos, persevere, think fundamentally, moonlight, and be arrogant. These recommendations are not in the same order as the alphabetical order of names, and the following commentaries are my sole responsibility.

EXPLOIT LUCK THROUGH OBSERVATION

This advice is a variation on the aphorism of Pasteur: Luck favors the prepared mind. A lot has been said about luck, chance discovery, or serendipity. It seems that the creative act cannot be planned consciously, so it appears accidental. But most often the *eureka* is preceded by months or years of study, inquiry, discussion, silent reflection, and meticulous experimentation. The deeper the thought, the more personal it is. Brainstorming has its uses in the research enterprise, as expressed by the well-known statement, discussion is the lifeblood of physics. Nevertheless the creative research climate must encourage meditation rather then agitation. Pressure does not hasten gestation of ideas. Yet, agitation and pressure have their roles in the development stages following creative research, but that is another story.

The *eureka* is not possible without observation. The mind must be prepared. Giulio Natta discovered isotactic polypropylene because he had the past expertise of X-ray diffraction that permitted him to recognize the ordered structure of his Nobel award-winning polymer. Without *quantitative* observation, there is no science. The creative research climate rewards prepared minds.

OPTIMIZE CHAOS

This sounds like an oxymoron, and it is one. In 1967, Herman Pines wrote in *Science* about Vladimir Ipatieff, the Russian chemist who left his imprint on the Universal Oil Products Company:

> Ipatieff was a general in the artillery of the czar. As befitted an officer at the time, he was a consummate horseman. Applying this skill to human relations, he used to say: "Give the subordinates enough rein, but let them know who the master is."

Another way to express the idea is to paraphrase Thomas Paine. The best research management is the least research management. Or, according to Paul Janssen, founder and chief of Janssen Pharmaceutica, research management should adapt its goals to the ability of the research staff to meet these goals. This is another way to optimize chaos. Who is a good research manager? A Ph.D. is neither a necessary nor a sufficient condition for a research manager. Many Ph.D.'s do not learn to *do* research: They may have been used only as a pair of hands by their thesis adviser. This is well recognized by the need for an assistant professor to demonstrate ability to do independent research before promotion to tenure rank.

The quality of a research laboratory is no better than that of its director. The director manages things, not people. By shrewd allocations of human and material resources, the research director creates a climate in which people believe that they are better than they really are. This climate is that of optimized chaos.

PERSEVERE

Nature is always more complex than it first seems. It always takes much more time than anticipated to solve a scientific problem. Even a patient director becomes impatient. The researcher needs courage, sometimes beyond the call of duty, to persevere against heavy odds. Loyalty commands to fight the establishment. Perseverance requires character. Faith must overcome expediency.

Yet, if perseverance means to stay on course, unforeseen results may demand a sudden sharp shift of direction to follow new leads. Managers do not like this either; they are committed to an orderly battle plan and hate the hazards of guerrilla warfare.

Toleration of perseverance and the flexibility to shift course are other expressions of the art of optimizing chaos. Academe provides a very opportunistic climate for science. Although research proposals fund specific problems, it is understood that the principal investigator is free to switch course on the basis of new information. Yet, the graduate student in the third year of doctoral work is better advised to persevere. These situations illustrate again the delicate balance between order and freedom in the research enterprise.

THINK FUNDAMENTALLY

To think fundamentally gives the scientist a chance to understand, or at least to understand better, the problem under study. But understanding means different things to different scientists. To the physicist, it means the ability to predict; to the chemist, it provides a way to explain. The engineer understands when design is at hand. At the lowest level, understanding is related to an orderly description of the facts.

Fundamental thinking in science must be quantitative. A non-quantitative activity is not science. All too often, the chemist thinks in terms of qualitative mechanisms. Mechanistic obsession is fun, at least to some, but it rarely helps, except when it leads to the invention of chemical reactions, as argued by Derek Barton (1990), or to the discovery of new molecules, as explained by Roald Hoffmann (1990).

No matter how it is done, thinking is always hard. It takes time and detracts from getting things done. Time to think is not the favorite allocation of managers. In fact, the scientist may well be forced to borrow thinking time from free time, following the imperative discussed in the next section.

MOONLIGHT

In the world of management, employees are nonexempt or exempt, depending on whether they get paid for overtime or not. But a creative scientist, like a creative artist, does not sell his or her time. The very idea of a creative scientist filling out time sheets is ludicrous. Fascination in science cannot be turned on or off on demand. The creative mind continues to wander while the body eats, exercises, or sleeps. Interruptions in or out of the laboratory, the library, the office, or the home study may ruin a promising effort. Hence the well-established practice of moonlighting. Do what you need to do to keep the wolf away from the door, the wolf being your thesis adviser, team leader, or laboratory director. Then use the rest of your time, perhaps at night or on the weekends, to do what you really want to do. All of us who have done science know how to moonlight effectively. So many scientific books are prefaced by remarks such as "I thank my spouse, who tolerated my awful antics while this book was being written." I acknowledge that some of the most creative results to come out of my laboratory were obtained by graduate students or postdoctoral assistants in the absence of my instructions or even occasionally against them.

The creative mind has a vision. And vision, following Jonathan Swift, is the art of seeing things invisible. To try to explain a vision can get one into serious trouble, as Joan of Arc found out. It is better to moonlight until things become visible.

By now, the reader must have noted many connections between the various admonitions to succeed in the research enterprise: the director must optimize chaos where the scientists count on luck, think endlessly, pursue seemingly hopeless avenues on or off the workplace, and sometimes act in a rather arrogant manner.

BE ARROGANT

At the end of their famous book on *The Conservation of Orbital Symmetry*, Woodward and Hoffmann (1970) consider violations of the rule that bears their name. They write, "Violations. There are none! Nor can violations be expected of so fundamental a principle of maximum bonding."

I define arrogance as the utterance of such a statement. To find something new is always risky. Is it really going to work? Is it right? The discoverer needs a champion. The best champion is the discoverer. This applies to new concepts, compositions of matter, processes, or technologies. Being arrogant does not mean acting as a bully. On the contrary, the successful inventor or entrepreneur combines arrogance with charm, being sometimes forceful in some sort of a shy manner. In any event, I far prefer the word arrogance to salesmanship, although they are not unrelated in the present context.

A SEVENTH KEY TO SUCCESS IN SCIENCE AND TECHNOLOGY

It would be foolhardy to claim that there are only six ways to success in the long road from science to an economically rewarding technology. To lengthen my list, I interrogated a man who has walked along this road not once but several times, Ralph Landau, to whom this essay is dedicated. What had been for him a guiding principle distinct from the six others that I already knew about? Ralph did not answer, but Claire Landau said, "Timing." Later she added, "And daring." So I propose a seventh pillar of success in the research enterprise: timing and daring. They seem inseparable, as timing, in the face of incomplete information, requires the daring of a risk taker. Albert Einstein talked about the value of timing in science, and daring he certainly was.

In conclusion, I have tried to comment on a number of attitudes that lead to success in the research enterprise of science and its applications. I have attempted to remain as general as possible, covering the gamut of activities from discovery to innovation. I fully realize that there are differences of style and substance as an idea moves from its inception to the industrial plant. But it is important to reemphasize that the creative and successful scientist-technologist-engineer is not just another professional and must be treated accordingly in hiring, promoting, rewarding, and if necessary, retiring or dehiring. Otherwise, the best creative minds will avoid science and its applications and will choose other pursuits.

In summary, let me rephrase some of my remarks. The best research management is the least management, alert to exploiting luck, avoiding a yo-yo style of starts and stops, convinced that *good science* always pays off. It is the lone guy in a corner of the laboratory who is likely to be the discoverer or inventor. But it is the arrogant guy who is likely to be the innovator with the seventh sense of timing and daring.

REFERENCES

Barton, D. 1990. The invention of chemical relations. Aldrichimica Acta 23:3.

Hoffman, R. 1990. Marginalia: Creation and discovery. American Scientist 78:14.

Pines, H. 1967. Man scientist. Science 157:167.

Woodward, R. B., and R. Hoffman. 1970. The Conservation of Orbital Symmetry. Weinheim: Verlag Chemie.

Michel Boudart is currently William M. Keck Professor of Chemical Engineering, Department of Chemical Engineering, at Stanford University. Born in Brussels, Belgium, he graduated from the University of Louvain with a B.S. degree (Candidate Ingenieur) and an M.S. degree (Ingenieur Civil Chimiste) a few years later. He received his Ph.D. degree in chemistry from Princeton University. After graduation he remained at Princeton as a research associate in the Forrestal Research Center. He became assistant to the director of Project SQUID and then assistant professor and, shortly thereafter, associate professor in the Department of Chemical Engineering. After a three-year stay at the University of California, Berkeley, as professor of chemical engineering, he became professor of chemical engineering and chemistry at Stanford University. Dr. Boudart is a founder of Catalytica, Inc. and is the author of numerous scientific papers on kinetics and catalysis. He is a member of the National Academies of Sciences and Engineering and a foreign member of the Belgian Royal Academy.

Challenges to Agricultural Research in the Twenty-first Century

Vernon W. Ruttan

In these remarks I will discuss some of the challenges facing the global agricultural research systems as we move into the first decades of the next century. Before doing so, however, I would like to first place my remarks within the intellectual climate that has conditioned our thinking about the relationships among environmental, technological, and institutional change during the second half of the twentieth century. I will then turn to some of the sources of stress from scientific, populist, and ideological sources that have buffeted the agricultural research community over the last several decades. Finally, I will report on some of the findings for research that have emerged from several recent "consultations" that I have organized around the issues of (a) biological and technical constraints on crop and animal productivity; and (b) resource and environmental constraints on sustainable growth in agricultural production.

TECHNOLOGY, INSTITUTIONS, AND THE ENVIRONMENT

The research that is conducted in our universities, research institutes, and our agricultural experiment stations is valued primarily for its contributions to technical and institutional change. The demand for advances in knowledge in the social sciences and humanities, and in related professional fields, is derived primarily from the demand for institutional change and more effective institutional performance.

There are several ways of characterizing the significance of technical change. It permits the substitution of knowledge for resources; it permits the substitution of more abundant for less abundant resources; and it releases the constraints on growth imposed by inelastic resources supplies. But technical change is itself the product of institutional innovation. Whitehead insisted that the greatest invention of the nineteenth century was the institutionalization of the process of invention—the invention of the research university, the industrial research laboratory, and the agricultural experiment station. One effect of the lag in institutional innovations needed to achieve an incentive-compatible institutional infrastructure—institutions capable of achieving compatibility between individual, organizational, and social objectives—is that the by-products of technical change, what the resource economists refer to as residuals, are now filling the landscape with garbage and the earth, water, and atmosphere with chemicals.

I am prepared to insist that the contributions of advances in natural and social science knowledge to technical and institutional change have enabled modern society to achieve a more productive and better balanced relationship to the natural world than was achieved in the ancient civilizations or in earlier stages of Western industrial civilization. But the relationship between advances in knowledge, resource use, and human well-being continues to be uneasy. We are, for example, in the midst of the third wave of social concern about the relationships between natural resources and the sustainability of improvements in human well-being since World War II—and the fifth since Malthus.

The *first* postwar wave of concern, in the late 1940s and early 1950s, focused primarily on the quantitative relations between resource availability and growth—the adequacy of land, water, energy, and other natural resources to sustain growth. The reports of the President's Water Resources Policy Commission and the President's Materials Policy Commission were the landmarks of the early postwar resource assessment studies generated by this wave of concern. The response to this first wave of concern was technical change. A stretch of high prices has not yet failed to induce the new knowledge and new technologies needed to locate new deposits, promote substitution, and enhance productivity. If the Materials Policy Commission report were writing today, it would have to conclude that there has been abundant evidence "of the nonevident becoming evident; the expensive, cheap; and the inaccessible, accessible."

The *second* wave of concern occurred in late 1960s and early 1970s. In this second wave, the earlier concern with the potential "limits to growth" imposed by natural resource scarcity was supplemented by concern about the capacity of the environment to assimilate the multiple forms of pollution generated by growth. An intense conflict was emerging between the two major sources of demand for environmental services. One was the rising demand for environmental assimilation of residuals derived from growth in commodity production and consumption—asbestos in our insulation, pesticides in our food, smog in the air, and radioactive wastes in the biosphere. The second was the rapid growth in consumer demand for environmental amenities—for direct consumption of environmental services—arising out of rapid growth in per capita income and high income elasticity of demand for such environmental services as access to natural environments and freedom from pollution and congestion. One response to these concerns, still incomplete, was the design of local incentive-compatible institutions designed to force individual firms and other organizations to bear the costs arising from the externalities generated by commodity production.

Since the mid-1980s these two earlier concerns have been supplemented by a *third*. These more recent concerns center around the implications for environmental quality, food production, and human health of a series of environmental changes that are occurring on a transnational scale—issues such as global warming, ozone depletion, acid rain, and others. The institutional innovations needed to respond to these concerns will be more difficult to design. They will, like the sources of change, need to be transnational. Experience with attempts to design incentive-compatible transnational regimes, such as the Law of the Sea Convention, or even the somewhat more successful Montreal Protocol on reduction of CFC emissions, suggests that the difficulty of resolving free-rider and distributional equity issues imposes a severe constraint on how rapidly effective transnational regimes to resolve these new environmental concerns can be put in place.

STRESS ON THE AGRICULTURAL RESEARCH SYSTEM

During the last century, American agriculture has made the transition from a natural resource-based industry to a science-based industry. During this period, rapid productivity growth enabled the agricultural sector to strengthen its position in world markets while simultaneously releasing much of its labor force to the nonfarm sectors of the economy. This is in sharp contrast to recent experience in the U.S. manufacturing sector. During the last decade and a half, lagging productivity growth in traditional and high-technology manufacturing has resulted in a loss of both jobs and competitive position in world markets.

The agricultural research community has taken considerable pride in its contribution to the remarkable economic performance of the agricultural sector over the last century (see Table 1). But this pride has been severely shaken. During the 1970s and early 1980s the closely articulated U.S. Department of Agriculture (USDA)–land-grant university research system was subject to considerable criticism from

TABLE 1 Average Annual Rates of Change (percentage per year) in Output, Inputs, and Productivity in U.S. Agriculture, 1870-1986.

Item	1870–1900	1900–1925	1925–1950	1950–1965	1965–1980	1980–1989
Farm output	2.9	0.9	1.6	1.7	2.0	0.4
Total inputs	1.9	1.1	0.2	–0.4	0.4	–2.6
Total productivity	1.0	–0.2	1.3	2.2	1.6	2.8
Labor inputs[a]	1.6	0.5	–1.7	–4.8	–2.8	–3.0
Labor productivity	1.3	0.4	3.3	6.6	4.9	3.3
Land inputs[b]	3.1	0.8	0.1	–0.9	0.8	–1.3
Land productivity[c]	–0.2	0.0	1.4	2.6	1.2	1.7

SOURCES: U.S. Department of Agriculture, *Economics Indicators of the Farm Sector: Production and Efficiency Statistics*, (Washington, D.C., USDA, Economic Research Service, January 1987); U.S. Department of Agriculture, *Changes in Farm Production and Efficiency* (Washington, D.C.: 1979); and D. D. Durost and G. T. Barton, *Changing Sources of Farm Output* (Washington, D.C.: USDA Production Research Report No. 36, February 1960). Data are three-year average centered on the year shown for 1925, 1950, 1965, and 1986.

[a]Number of workers, 1870-1910; worker-hour basis, 1910-1971.
[b]Cropland use for crops, including crop failures and cultivated summer fellow.
[c]Total output (not just crop output) per acre of cropland.

NOTE: The 1988 *Production and Efficiency Statistics* report will introduce a new Divisa-based total (or multifactor) productivity index. This may result in some changes in productivity growth rates.

populist, scientific, ideological perspectives. At the risk of some oversimplification, it may be useful to characterize these criticisms along the following lines.

The criticism directed toward agricultural research by the general science community was that agricultural research was not "good science." A central element in this negative perception of agricultural research was that it has been funded primarily through institutional support rather than through competitive grants. A second element is that a relatively high share of agricultural research has been directed toward technology development. While generally conceding that the investment in agricultural research has paid high social dividends in the past, there was concern that the system was losing its capacity to make comparable contributions in the future.

An ideological criticism that emerged with particular force in the Office of Science and Technology Policy in the early 1980s, was a perception that public research support should be confined to the basic sciences and that the private sector should be primarily responsible for applied research. The proponents of this view tend to avoid questions of the articulation or synergy between basic and applied research. There was also an even greater reluctance to address the problem of how to ensure research performance in those areas of technology development where private incentives are inadequate to generate an economically or socially optimum level of research.

The populist critics have viewed agricultural research, and the technology that has been generated by agricultural research, as responsible for the displacement of small farms and farm workers, as a source of the decline of rural communities, as a cause of deterioration in the quality and safety of food, and as an assault on the quality of the environment. Thus, in the populist view, agricultural research was regarded as a powerful instrument of technical and social change that has been captured by organized agribusiness and has misdirected its energies against the people and the institutions that it was designed to serve. During the early and mid-1980s, a global recession and the rising value of the dollar combined to dampen the demand for U.S. agricultural commodities abroad. High interest rates, associated first with inflation and later with massive federal borrowing, imposed severe financial burdens on farmers and their suppliers. These combined to force a decline in farm commodity prices, severe deflation in land values, and a financial crisis for many farmers. Some critics suggested a moratorium on agricultural research and technology development. Others called for the transfer of resources from research directed to productivity enhancement to cost reduction—apparently without realizing that these were opposite sides of the same coin. State agricultural experiment stations were urged to withdraw from efforts supported by the U.S. Agency for International Development to strengthen national research systems in developing countries.

By the end of the 1980s new pressures were being brought to bear on the U.S. agricultural research system. Concerns about the impact of agricultural intensification widened. In the 1970s these concerns had initially focused on the effects of pesticides and nonpoint sources of pollution on natural environments and on the safety of farm workers and consumers. During the 1980s concerns about the effects of more intensive agricultural production on (a) resource degradation through erosion, salinization, and depletion of groundwater; and (b) the quality of surface and groundwater through runoff and leaching of plant nutrients and pesticides intensified. Terms that had been introduced by the populist critics of agricultural research—such as alternative, low-input, regenerative, and sustainable agriculture—began to enter the vocabulary of those responsible for allocating resources for agricultural research. After an initial period of resistance, some leaders of the agricultural research community moved to embrace this new set of concerns. The recently issued report by the National Research Council Board on Agriculture on *Alternative Agriculture* has been viewed as a landmark in this conversion. In my judgment, this report is more appropriately viewed as a political document designed to capture the initiative from the populist critics of institutionalized agricultural research.

The changes that I have described can be put in a somewhat broader context. During the last two decades, the agricultural research system has been attempting to respond to a new set of demands—and opportunities—resulting from populist, scientific, and ideological challenges in an environment in which its access to economic and political resources has been declining. Federal agricultural research funding and performance have stagnated since the late 1960s (Figures 1 and 2). By the late 1980s the USDA provided only about 16 percent of total federal support for academic basic research in plant biology. An increasing share of the USDA Agricultural Research Service research was supported by transfers from other agencies. Modest growth of

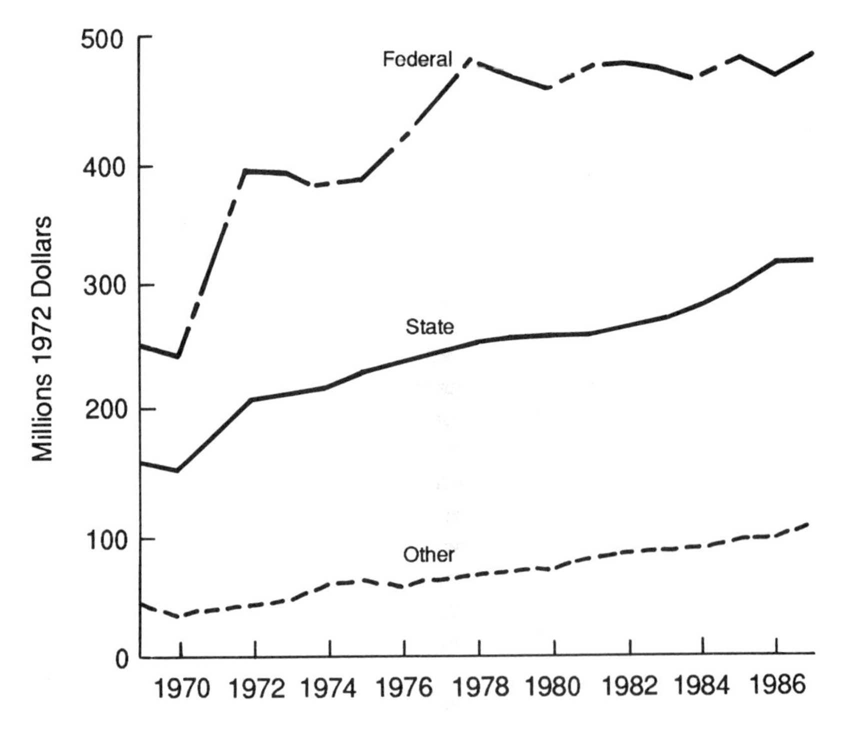

FIGURE 1 Sources of funding for government R&D. SOURCE: Pray (1989).

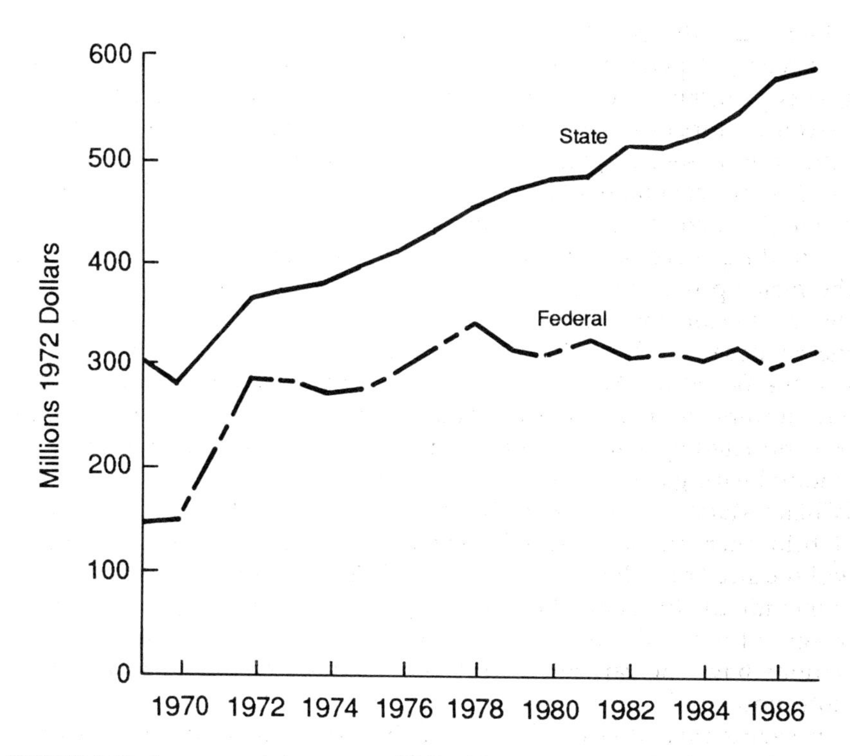

FIGURE 2 Performance of government R&D. SOURCE: Pray (1989).

state support for agricultural research has been sufficient to slightly more than offset the decline in federal support. Private sector agricultural research has risen from a level roughly equivalent to the level of the USDA–land-grant system in the mid-1960s to close to 60 percent of the total in the mid-1980s. However, there is some evidence of a decline in private sector research, both in the newer areas of biotechnology and in the more traditional areas of biological technology such as plant breeding, since the mid-1980s.

In an attempt to reverse the stagnation in public support for agricultural research, the National Research Council Board on Agriculture issued a report in the fall of 1989 calling for an increase in funding of $500 million for the competitive research grants program administered by the USDA. The program would support research in public and private universities, the USDA research agencies, and other research agencies of the state and federal government. Despite the boldness of the proposal, it would not be surprising to see an increase in funding of the magnitude achieved over a five-year period—by which time the purchasing power of the increase may have been reduced by 25 percent as a result of inflation. If an expanded competitive grants program, even of the size proposed, is to make more than a marginal impact on agricultural research capacity, the support for other public sector agricultural research will have to increase from a level of approximately $2.0 billion in 1986 to the $2.5 to $3.0 billion range (in current dollars) by the mid-1990s.

Funding increases in the range discussed above will not resolve the major problems facing the agricultural research system. One of the most critical is the issue of who will do the research. In framing the proposal for the $500 million increase in competitive grant funding, the Board on Agriculture was unable to avoid a conclusion that had become obvious a decade ago—most of the new research would be conducted by a subset of elite institutions. In testimony presented at joint hearings by the Senate and House Agriculture Subcommittees, William Marshall, who heads the Microbial Genetics Division of Pioneer Hybrid International, insisted that the viability of agricultural research will require "broadening of the scientific base in agricultural research to include the fundamental sciences outside of the land-grant colleges of agriculture." Furthermore, "no more than 15 of our 57 experiment stations have the capability to do significant amounts of research in biotechnology."

It seems clear that by the end of the first decade of the next century, the agricultural research landscape will look much different than it does today. Moreover, pressures for the revision of research priorities arising from scientific, societal, and environmental change will not abate.

BIOLOGICAL AND TECHNICAL CONSTRAINTS

During the last six months, I have had an opportunity, with support from the Rockefeller Foundation, to organize a series of small "consultations" on the agricultural research priorities that might be expected to emerge as we move into the early decades of the next century. The first of these consultations was organized around the topic "Biological and Technical Constraints on Crop and Animal Productivity," and the second around the issues of "Resource and Environmental Constraints on Sustainable Growth in Agricultural Production." Although the consultations were not confined to domestic priorities, the issues and conclusions were quite relevant to U.S. agricultural research policy.

Those familiar with the evidence on long-term declines in agricultural commodity prices (Figures 3 and 4) or with media attention to

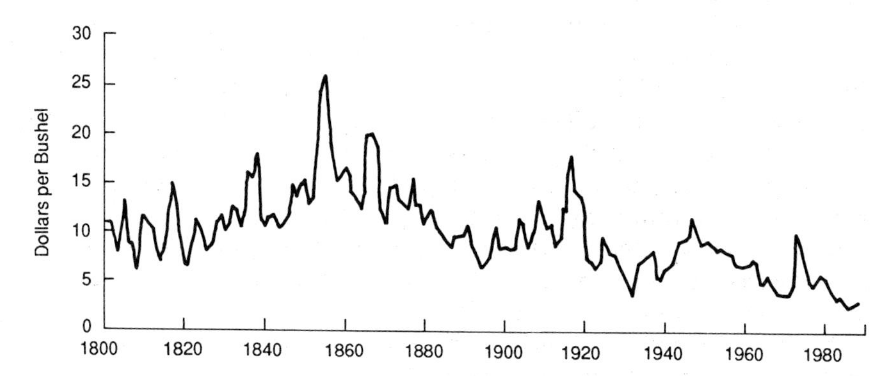

FIGURE 3 Real wheat prices since 1800. SOURCE: Edwards (1988).

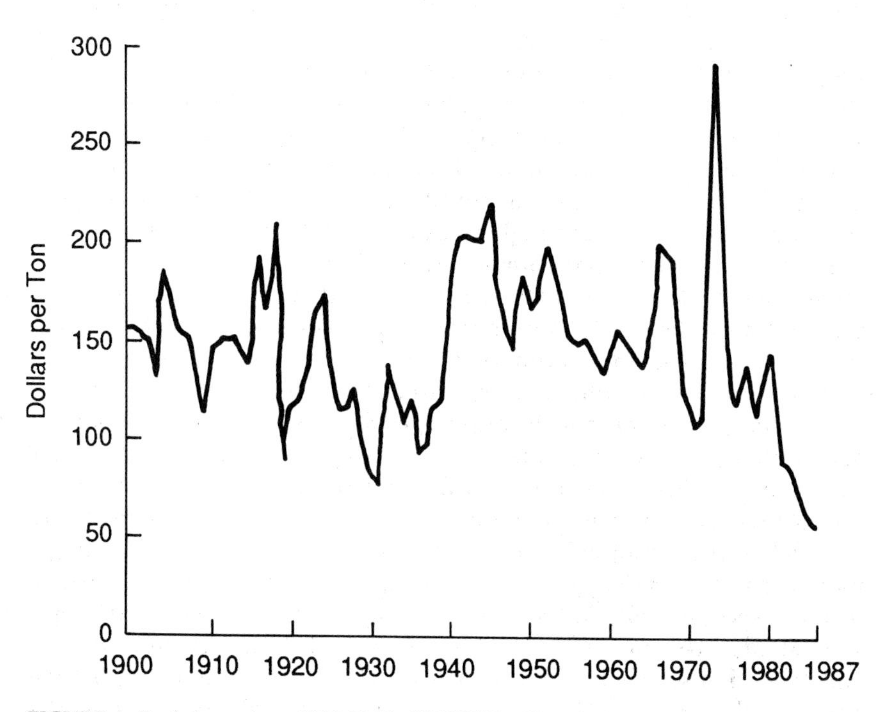

FIGURE 4 Real rice prices, 1900-1987. SOURCE: Pingali (1988).

the "new biotechnology" may find it difficult to comprehend why anyone should be concerned about the possibilities of a lag in either agricultural production or productivity over the next several decades. Let me justify my concern with just four observations: (1) the yields obtained on maximum yield trials at the International Rice Research Institute are no higher today than they were in the mid-1960s; (2) maize yields in the United States continue to increase at about one bushel per year—but this is a much smaller rate of increase than 30 years ago; (3) the projected impact of biotechnology on agricultural production continues to be postponed—benefits expected in this decade are now expected in the next; and (4) national agricultural research capacity has weakened in a number of debt-plagued developing countries and in Eastern Europe and the USSR.

Let me now turn to some major conclusions from the consultation on biological and technical constraints.

Advances in conventional technology will remain the primary source of growth in crop and animal production over the next quarter century. Almost all future increases in agricultural production must come from further intensification of production on land that is now devoted to crop and livestock production. Until well into the second decade of the next century, the necessary gains in crop and animal productivity will continue to be generated by improvements resulting from conventional plant and animal breeding and from more intensive and efficient use of technical inputs, including chemical fertilizers, pest-control chemicals, and higher quality animal feeds. The productivity gains from conventional sources are likely to come in smaller increments than in the past. If they are to be realized, higher plant populations per unit area, new tillage practices, improved pest and disease control, more precise application of plant nutrients, and advances in soil and water management will be required. Gains from these sources will be crop, animal, and location specific. They will require closer articulation between the suppliers and users of knowledge and new technology. These sources of productivity gains will be extremely knowledge and information intensive. If they are to be realized, research and technology transfer efforts in information and management technology must become increasingly important sources of growth in crop and animal productivity. In the short run, that is, the next several decades, no other sources of growth in production will be adequate to meet the demands that will arise from growth in population and income, and be placed on agricultural production in either the developed or developing countries. Both national and international agricultural research systems will need to increase the proportion of research resources devoted to improvement of agronomic practice relative to plant breeding.

Advances in conventional technology will be inadequate to sustain the demands that will be placed on agriculture in the second decade of the next century and beyond. Advances in crop yields have come about primarily by increasing the ratio of grain to straw rather than by increasing total dry matter production. Advances in animal feed efficiency have come by decreasing the proportion of consumed feed that is devoted to animal maintenance and increasing the proportion that produces usable animal products. There are severe physiological constraints to continued improvement along these conventional paths. These constraints are most severe in those areas that have already achieved the highest levels of productivity—as in Western Europe, North America, and parts of East Asia.

The impact of these constraints can be measured in terms of declining incremental response to energy inputs—in the form of reductions in both the incremental yield increases from higher levels of fertilizer application, and the incremental savings in labor inputs from the use of larger and more powerful mechanical equipment. One consequence is that in countries that have achieved the highest levels of output per hectare or per animal unit, an increasing share of both public and private sector research budgets is being devoted to maintenance research—the research needed to sustain existing productivity levels. A decline in the incremental returns to agricultural research would impose a higher priority on efficiency in the organization of research and on the allocation of research resources.

A reorientation of agricultural research will be necessary to realize the opportunities for technical change being opened up by advances in microbiology and biochemistry. Advances in basic science, particularly in molecular biology and biochemistry, continue to open up new possibilities for supplementing traditional sources of plant and animal productivity growth. Possibilities range from the transfer of growth hormones into fish to the conversion of lignocellulose into edible plant and animal products.

The realization of these possibilities will require a reorganization in the performance of agricultural research. An increasing share of the new knowledge generated by research will reach producers in the form of proprietary products or services. This means that the incentives exist to draw substantially more private sector resources into agricultural research. Public sector research organization increasingly will have to move from a "little science" to a "big science" mode of organization. Examples include the Rockefeller Foundation-sponsored collaborative research program on the biotechnology of rice and the University of Minnesota program on the biotechnology of maize. In the absence of more focused research efforts, it seems likely that the promised gains in agricultural productivity from biotechnology will continue to recede in the future.

Efforts to institutionalize agricultural research capacity in developing countries must be intensified. Crop and animal productivity levels in most developing countries remain well below the levels that are potentially feasible. Access to the conventional sources of productivity growth—from advances in plant breeding, agronomy, and soil and water management will require the institutionalization of substantial agricultural research capacity. In a large number of developing countries this capacity is just beginning to be put in place. A number of countries that experienced substantial growth in capacity during the 1960s and 1970s have experienced an erosion of capacity in the 1980s. Even a relatively small country, producing a limited range of commodities under a limited range of agro-climatic conditions, will require a cadre of 250-300 agricultural scientists. Countries that do not acquire adequate agricultural research capacity will not be able to meet the demands placed on their farmers as a result of growth in population and income. Research systems that do not generate resource and productivity enhancing capacity will fail to sustain public support.

There are substantial possibilities for developing sustainable agricultural production systems in a number of fragile resource areas. Research in the tropical rain forest areas of Latin America and in the semiarid tropics of Africa suggests the possibility of developing sustainable agricultural systems with substantially enhanced productivity. It is unlikely, and perhaps undesirable, that these areas become important components of the global food supply system. But enhanced productivity is important to the people who live in these areas now and in the future. It is important that the research investment in soil and water management and in farming systems be intensified in these areas.

Over the long run, energy and mineral nutrition can be expected to emerge as increasingly serious constraints on agricultural production. During the last century, technical change has been directed along alternative paths in different countries by their relative resource endowments. Countries where land was relatively scarce or expensive, such as Japan, placed an emphasis on biological technology—in effect, inventing around the land resource constraint. Countries where labor was relatively scarce or expensive, such as the United States, placed greater emphasis on advancing mechanical technology—in effect inventing around the labor constraint. Over the next half century, energy derived from liquid fuels is likely to become a serious constraint. It is also possible that the reserves of phosphate raw material will decline to levels that will result in much higher relative prices for phosphate fertilizer. It is likely that it will be necessary to allocate substantial research resources to invent around these two constraints.

The rationalization of regulatory regimes will become an increasingly important factor in determining the profitability of research investments and international competitiveness in agricultural production. Incentives for private sector agricultural research appear to be quite sensitive to uncertainty about changes in regulatory regimes and the administration of regulations. Incentives for research and the potential gains from research investment are reduced when use of technology is restricted for reasons other than the assurance of health and safety. Consumers may press for regulation based on aesthetic concerns. Producers may press for regulation to protect themselves from domestic or international competition. Pressure to achieve greater consistency among national regulatory regimes is likely to become an increasingly important factor in international trade negotiations. It will be necessary to devote substantial research efforts to identifying and quantifying the scientific, technical, economic, and psychological information needed to rationalize regulatory regimes in the future.

A major effort to assemble and characterize available plant and animal genetic resources is essential to make the transition from the conventional biological technology of the twentieth century to a biotechnology-based agriculture for the twenty-first century. A major constraint in the development of a cost-effective strategy for collection and preservation of genetic resources is an adequate characterization of the materials in *in situ* locations and in *ex situ* collections. A genome mapping program for crop plants is essential if we are going to make effective use of the genetic engineering techniques that are available now and that will become available in the future.

Research on alternative crops and animals that can be introduced into production systems can become a useful source of growth in some areas. On a local or regional basis, the development and incorporation of minor cultivars and species could make important nutritional and economic contributions. It is unlikely that alternative crops or animals will emerge to substantially replace existing crop cultivars or animal species in production systems. It would be wishful thinking to expect any new developments as significant as the expansion of soybean production during the past half century.

There is a need to establish substantial basic biological research and training capacity in the tropical developing countries. A number of basic biological research agendas that are important for applied research and technology development in health and agriculture in the tropics receive, and are likely to continue to receive, inadequate attention in the temperate region developed countries. There is also a need for closer articulation between training in applied science and technology and training in basic biology. When such institutes are established, they should be more closely linked with existing universities than the series of agricultural research institutes established by the Consultative Group on International Agricultural Research.

RESOURCE AND ENVIRONMENTAL CONSTRAINTS

As we look even further into the next century, there is a growing concern, as noted earlier, about the impact of a series of resource and environmental constraints that may seriously impinge on our capacity to sustain growth in agricultural production. One set of concerns centers on the environmental effects of agricultural intensification. These include groundwater contamination from plant nutrients and pesticides, soil erosion and salinization, the growing resistance of insect pests and pathogens and weeds to present methods of control, and the contribution of agricultural production and land use changes to global climate change. The second set of concerns stems from the effects of industrial intensification of global climate change. It will be useful, before presenting some of the findings of the second consultation, to characterize our state of knowledge about global climate change.

There can no longer be any question that the accumulation of carbon dioxide (CO_2) and other greenhouse gases—principally methane (CH_4), nitrous oxide (N_2O), and chlorofluorocarbons (CFCs)—has set in motion a process that will result in some rise in global average surface temperatures over the next 30-60 years. There is substantial disagreement about whether warming due to greenhouse gases has already been detected. And there continues to be great uncertainty about the increases in temperature that can be expected to occur at any particular date or location in the future.

Most carbon dioxide emissions come from fossil fuel consumption. Carbon dioxide accounts for roughly half of radiative forcing (Figure 5). Biomass burning, cultivated soils, natural soils, and fertilizers account for close to half of nitrous oxide emissions. Most of the known sources of methane are a product of agricultural activities—principally enteric fermentation in ruminant animals, release of methane from rice production and other cultivated wetlands, and biomass burning. Estimates of nitrous oxide and methane sources have a very fragile empirical base. Nevertheless, it appears that agriculture and related land use could account for somewhere in the neighborhood of 25

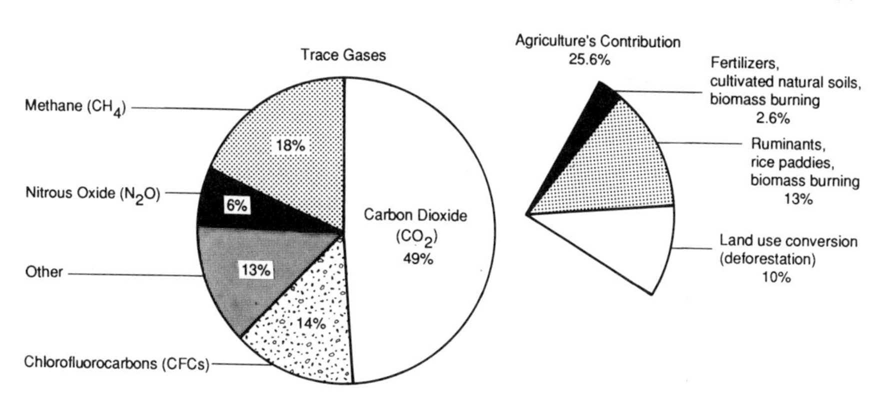

FIGURE 5 Contributions to increases in radiative forcing in the 1990s.
SOURCE: Reilly and Bucklin (1989).

percent of radiative forcing. On a regional basis the United States contributes about 20 percent of radiative forcing by all greenhouse gases while Western and Eastern Europe and the USSR contribute about 30 percent. In the near future, contributions to radiative forcing from the Third World will exceed those from the Organization for Economic Cooperation and Development and what used to be called the centrally planned economies.

Several participants in the second consultation characterized the alternative policy approaches to the threat of global warming as *preventionist* and *adaptionist*. It seems clear that a preventionist approach could involve about five policy options. They include reduction in fossil fuel use, or capture of CO_2 emissions at the point of fossil fuel combustion, reduction in the intensity of agricultural production, reduction of biomass burning, expansion of biomass production, and energy conservation.

The simple enumeration of these policy options should be enough to introduce considerable caution about assuming that radiative forcing will be limited to present levels. Let me be more specific. Fossil fuel use will be driven, on the demand side, largely by the rate of economic growth in the Third World and by improvements in energy efficiency in the developed and the centrally planned economies. On the supply side, it will be constrained by the rate at which alternative energy sources will be substituted for fossil fuels. Of these only energy efficiency and conservation are likely to make any significant contribution over the next generation. The speed with which it will occur will be limited by the pace of capital replacement. Any hope of significant reversal of agricultural intensification, reduction in biomass burning, or increase in biomass absorption is unlikely to be realized within the next generation. The institutional infrastructure or institutional resources that would be required do not exist and will not be put in place rapidly enough to make a significant difference.

The possibilities for energy conservation make it fairly easy to be cautiously optimistic about endorsing a preventionist approach in dealing with the industrial sources of climate forcing—at least in the currently industrialized countries. I see little alternative, however, to an adaptionist approach in attempting to assess how agricultural research portfolios should respond to the implications of global climate change. It also forces me to agree that we will not be able to rely on a technological fix to the global warming problem. The fixes, whether driven by preventionist or adaptionist strategies, must be both technological and institutional. An adaptionist strategy for agriculture implies moving as rapidly as possible to design and put in place the institutions needed to remove the constraints that intensification of agricultural production is currently imposing on sustainable increases in agricultural production. I am referring, for example, to the policies and institutions needed to rationalize water use in the western United States or to deal with groundwater management (including contamination) in both developed and developing countries. If we are successful in putting in place such policies and institutions, we will then be in a better position to respond to the more uncertain changes that will emerge as a result of future global climate change.

Let me now turn to some of the research implications that emerged from the consultation.

A major research program on incentive compatible institutional design should be initiated. The first research priority is to initiate a large-scale program of research on the design of institutions capable of implementing incentive-compatible resource management policies and programs. By incentive-compatible institutions I mean institutions capable of achieving compatibility among individual, organizational, and social objectives. A major source of the global warming and environmental pollution problem is the direct result of the operation of institutions that induce behavior by individuals, and public agencies that are not compatible with societal development—some might say survival—goals. In the absence of more efficient incentive-compatible institutional design, the transaction costs involved in *ad hoc* approaches are likely to be enormous. Substantial basic research will be required to support a successful program of applied research and institutional design.

A serious effort to develop alternative land use, farming systems, and food systems scenarios for the twenty-first century should be initiated. A clearer picture of the demands that are likely to be placed on agriculture over the next century and of the ways in which agricultural systems might be able to meet such demands has yet to be produced. World population could rise from the present 5 billion level to the 10–20 billion range. The demands that will be placed on agriculture will also depend on the rate of growth of income—particularly in the poor countries where consumers spend a relatively large share of income growth on subsistence—food, clothing, and housing. The resources and technology that will be used to increase agricultural production by a multiple of 3–6 will depend on both the constraints on resource availability that are likely to emerge and the rate of advance in knowledge. Advances in knowledge can permit the substitution of more abundance for increasingly scarce resources and reduce the resource constraints on commodity production. Past studies of potential climate change effects on agriculture have given insufficient attention to adaptive change in nonclimate parameters. But application of advances in biological and chemical technology, which substitute knowledge for land, and advances in mechanical and engineering technology, which substitute knowledge for labor, have in the past been driven by increasingly favorable access to energy resources—by declining prices of energy. It is not unreasonable to anticipate that there will be strong incentives, by the early decades of the next century, to improve energy efficiency in production and use of agricultural products. Particular attention should be given to alternative and competing uses of land. Land use transformation, from forest to agriculture, contributes to radiative forcing through release of CO_2 and methane into the atmosphere. Conversion of low-intensity agricultural systems to forest has been proposed as a method of absorbing CO_2. There will also be increasing demands on land use for watershed protection and biomass energy production.

The capacity to monitor the agricultural sources and impacts of environmental change should be strengthened. It is a matter of serious concern that only in the last decade and a half has it been possible to estimate the magnitude of soil loss in the United States and its effects on agricultural productivity. Even rudimentary data on soil loss are almost completely unavailable in most developing countries. The same point holds, with even greater force, for groundwater pollution, salinization, species loss, and other areas. It is time to design the elements of a comprehensive agriculturally related resource-monitoring system and to establish priorities for implementation. Data on the effects of environmental change on the health of individuals and communities is even less adequate. The monitoring effort should include a major focus on the effects of environmental change on human populations.

Lack of firm knowledge about the contribution of agricultural practices to the methane and nitrous oxide sources of greenhouse forcing was mentioned at numerous times during the consultation. Much closer collaboration between production-oriented agricultural scientists, ecological trained biological scientists, and the physical scientists that have been traditionally concerned with global climate change is essential. This effort should be explicitly linked with the monitoring efforts currently being pursued under the auspices of the International Geosphere-Biosphere Programs.

The design of technologies and institutions to achieve more efficient management of surface and groundwater resources will become increasingly important. During the twenty-first century water resources will become an increasingly serious constraint on agricultural production, which is already a major source of decline in the quality of both ground and surface water. Limited access to a clean and uncontaminated water supply is a major contributor to disease and poor health in many parts of the developing world and in the centrally planned economies. Global climate change can be expected to have a major differential impact on the water availability, water demand, erosion, salinization, and flooding. The development and introduction of technologies and management systems that enhance water use efficiency represent a high priority both because of short- and intermediate-run constraints on water availability and the longer-run possibility of seasonal and geographical shifts in water availability. The identification, breeding, and introduction of water-efficient crops for dryland and saline environments is potentially an important aspect of achieving greater water-use efficiency.

The modeling of the sources and impacts of climate change must become more sophisticated. One of the problems with both the physical and economic modeling efforts is that they have tended to be excessively resistant to advances in micro-level knowledge, including failure to take into consideration climate change response possibilities from agricultural research and the response behavior of decision-making units such as governments, agricultural producers, and consumers.

Research on environmentally compatible farming systems should be intensified. In agriculture, as in the energy field, there are a number of technical and institutional innovations that could have both economic and environmental benefits. Among the technical possibilities is the design of new "third" or "fourth" generation chemical, biorational, and biological pest management technologies. Another is the design of land use technologies and institutes that will contribute to reduction of erosion, salinization, and groundwater pollution.

Immediate efforts should be made to reform agricultural commodity and income support policies. In both developed and developing countries, producers' decisions on land management, farming systems, and use of technical inputs (such as fertilizers and pesticides) are influenced by government interventions such as price supports and subsidies, programs to promote or limit production, and tax incentives and penalties. It is increasingly important that such interventions be designed to take into account the environmental consequences of decisions by land owners and producers induced by the interventions.

Alternative Food Systems. A food-system perspective should become an organizing principle for improvements in the performance of existing systems and for the design of new systems. The agricultural science community should be prepared, by the second quarter of the next century, to contribute to the design of alternative food systems. Many of these alternatives will include the use of plants other than the grain crops that now account for a major share of world feed and food production. Some of these alternatives will involve radical changes in food sources. One such system is based on lignocellulose—both for animal production and human consumption.

PERSPECTIVE

In this concluding section I return to the problem of whether the public agricultural research system will respond to the new challenges and opportunities of (a) releasing the biological and technical constraints on crop and animal productivity; (b) reducing the contribution of the agricultural sector to environmental degradation; and (c) enabling the agricultural sector to adapt to those environmental changes that emerge in response to the intensification of industrial production. Issues of both scientific and political capacity are involved.

Two decades of erosion in research capacity, particularly at the federal level, have left the research system in a weakened position to respond to either—let alone both—sets of concerns. The significance of this decline is reinforced by the even more rapid decline in research support and capacity in the other federal resource agencies and in the very limited support and capacity for mission-oriented research in the academic biological and environmental sciences.

The capacity of the agricultural research system to respond is also weakened by the political constraints within which it functions. The traditional agricultural research clientele—the organized commodity groups, elements of the agribusiness community, and the members of the Congress and the state legislatures who have significant agricultural constituencies—are capable of bring considerable pressure to bear to limit the transfer of resources necessary to respond to the environmental research agenda. They doubt, correctly in my view, the capacity of the private sector to replace the traditional production-oriented research conducted by the public sector. Yet, they have not demonstrated in recent years the political resources necessary to secure expanded funding, or even the funding necessary to prevent erosion of capacity needed to respond to the challenge of meeting the constraints on agricultural production.

BIBLIOGRAPHY

Abrahamson, Dean E. 1989. The Challenge of Global Warming, National Resources Defense Council. Washington, D.C.: Island Press.

Ausubel, Jesse H., and Hedy E. Sladovich. 1989. Technology and Environment. Washington, D.C.: National Academy Press.

Batie, Sandra S., "Sustainable Development: Challenges to the Profession of Agricultural Economics," American Journal of Agricultural Economies, December 1989.

Board on Agriculture, National Research Council. 1987. Agricultural Biotechnology: Strategies for National Competitiveness. Washington, D.C.: National Academy Press.

Board on Agriculture, National Research Council. 1989. Investing in Research: A Proposal to Strengthen the Agricultural Food and Environmental System, Washington, D.C.: National Academy Press.

Board on Agriculture, National Research Council. 1989. New Directions for Biosciences Research in Agriculture: High Reward Opportunities. Washington, D.C.: National Academy Press.

Committee on the Role of Alternative Farming Methods on Modern Production Agriculture, Board on Agriculture, National Research Council. 1989. Alternative Agriculture. Washington, D.C.: National Academy Press.

Edwards, Clark. 1988. Real prices received by farmers keep falling. Choices Fourth Quarter:22-23.

Keyworth, George A., II. 1984. Four years of Reagan science policy: Notable shifts in priorities. Science 224 (April 6): 9-13.

National Research Council. 1972. Report of the Committee on Research Advisory to the U.S. Department of Agriculture, National Technical Information Service, Springfield, Virginia.

Pingali, Prabhu. 1988. Intensification and diversification of Asian rice farming systems. International Rice Research Institute Agricultural Economics Paper 88-41, Los Banos, Laguna, Philippines.

Pray, Carl E. 1989. Research Policy for U.S. Food and Agriculture in the 1990s: R&D Trends, Problems, and Policy Instruments. Rutgers University Department of Agricultural Economics, New Brunswick, N.J., November 1989 (mimeo).

Reilly, John and Bucklin, Rhonda. 1989. Climate change and Agriculture. World Agriculture Situation and Outlook Report. Washington, D.C. USDA/ERX, WAS-55 June.

Ruttan, Vernon W. 1982. Agricultural Research Policy. Minneapolis, Minn.: University of Minnesota Press.

Ruttan, Vernon W. 1971. Technology and the environment. American Journal of Agricultural Economics 53 (December):707-717.

Ruttan, Vernon W. 1988. Sustainability is not enough. American Journal of Alternative Agriculture. Spring/Summer:128-130.

Ruttan, Vernon W., ed. 1989. Biological and Technical Constraints on Crop and Animal Productivity: Report on a Dialogue, Department of Agricultural and Applied Economics. University of Minnesota, St. Paul, December 1989.

Ruttan, Vernon W., ed. Resource and Environmental Constraints on Sustainable Growth in Agricultural Production, Department of Agricultural and Applied Economics, University of Minnesota, St. Paul, forthcoming.

Ruttan, Vernon W., and Carl Pray, eds. 1988. Policy for Agricultural Research. Boulder, Colo.: Westview Press.

VERNON D. RUTTAN is Regents' Professor in the Department of Agricultural and Applied Economics and the Department of Economics at the University of Minnesota, Minneapolis. Dr. Ruttan studied at Yale University where he received a B.A. degree and at the University of Chicago, where he received the M.A. and Ph.D. degrees. He spent the early portion of his career as an economist with the Tennessee Valley Authority and also as an assistant professor in the Department of Agricultural Economics at Purdue University. He has authored numerous articles in the field of agricultural economics and is a member of the National Academy of Sciences.

Innovation in the Chemical Processing Industries

Ralph Landau and Nathan Rosenberg

Chemicals and allied products (Standard Industrial Classification 28) is the high-tech sector about which the general public probably has the least knowledge. Yet, judged by criteria that are generally regarded as socially and economically worthwhile, this sector should be ranked at the top of the high-tech scale. A common criterion for "high tech" is an industry's expenditure upon research and development (R&D). Chemicals and allied products is at the very top when industries are ranked in terms of the share of total R&D that is actually financed by private funds. With respect to the composition of R&D expenditures, a far larger share of such expenditures in this sector consists of basic research, and basic research and applied research together represent a much greater share of total R&D than is the case in any other industrial sector (see Table 1). It is tempting to say that this sector has received so little public attention because its performance has, in certain respects at least, been so exemplary.

Clearly, chemicals and allied products have been heavily dependent upon the performance of scientific research. Having said that, it must be emphasized that such research is only the very beginning of the innovation process, and not the end of it. A laboratory breakthrough is, typically, very far from the availability of a commercializable product. Commercial success or failure in this industry, as in other industries, is largely a matter of what happens after a laboratory discovery. However significant the contribution of science to human welfare in general, the question of who will benefit most from specific innovations generated by science will depend on factors far removed from scientific research capability.

In chemicals, and especially organic chemicals, the development of

TABLE 1 Percentage Composition of the R&D Expenditures in the Six Industries of the U.S. Economy where R&D is Mostly Concentrated

	Chemicals and Allied Products			Nonelectrical Machinery			Electrical Machinery		
	BR	AR	D	BR	AR	D	BR	AR	D
1965	12.9	38.8	48.3	2.1	12.9	85.0	4.6	13.5	81.9
1966	18.0	39.0	48.0	2.1	13.0	85.0	4.0	12.0	84.0
1967	12.4	37.8	49.7	2.0	13.2	84.8	3.4	12.9	83.7
1968	12.5	37.1	50.4	1.9	12.8	85.2	3.3	13.7	83.0
1969	12.7	38.5	48.8	1.3	15.2	83.5	3.1	15.0	81.9
1970	12.0	38.7	49.3	1.3	15.2	83.5	3.3	15.0	81.7
1971	13.2	38.8	47.9	1.1	14.0	84.9	3.2	15.3	81.5
1972	12.9	39.5	47.6	1.2	13.6	85.2	2.9	16.3	80.8
1973	10.8	40.8	48.4	1.1	13.6	85.3	3.3	15.6	81.1
1974	11.3	39.8	48.9	1.0	13.0	86.0	3.3	15.6	81.1
1975	10.4	38.9	50.6	1.1	12.2	86.7	3.5	15.7	80.8
1976	10.1	41.0	48.9	1.6	11.3	87.1	2.9	17.4	79.8
1977	10.3	41.7	48.0	1.5	11.2	87.3	3.0	17.1	79.8
1978	n/a	n/a	n/a	n/a	n/a	n/a	n/a	na/	n/a
1979	9.1	41.6	49.3	1.3	13.1	85.5	3.0	15.4	81.6
1980	n/a	n/a	n/a	n/a	n/a	n/a	n/a	n/a	n/a
1981	10.1	42.5	47.4	1.9	18.4	79.7	2.7	17.0	80.3
1982	n/a	n/a	n/a	n/a	n/a	n/a	n/a	n/a	n/a
1983	46.0		54.0	1.4	14.5	84.0	3.1	16.5	80.4
1984	8.4	37.4	54.2	1.6	14.0	84.4	3.0	16.7	80.4

TABLE 1 continues

new products depends on the findings of scientific experiments performed at the laboratory level. The initial stages in the development of new polymers, for instance, depend on the laboratory combination of individual molecules (monomers) to form a single composite molecule (polymers). Depending on the length, the shape, and the chemical properties of the individual monomers, one may create materials with different chemical and physical properties, such as plastics, resins, synthetic rubber and fibers, films, and foams. Of course the role of laboratory research becomes relatively less important at later stages of the development process when chemical engineering becomes the fundamental discipline for transforming the bench-scale reactions to production on a full industrial manufacturing scale. Yet the particular nature of the products and the production processes in chemicals accounts for the significance of scientific research at the early stages of the innovation development cycle, which sets closer ties between science and production than is the case in other industrial realms.

TABLE 1 (continued)

	Automobiles and Other Transportation Equipment			Aeronautics and Missiles			Scientific and Professional Instruments		
	BR	AR	D	BR	AR	D	BR	AR	D
1965	3.0	97.0		1.4	14.3	84.3	n/a	n/a	n/a
1966	3.0	97.0		1.0	14.0	85.0	n/a	n/a	n/a
1967	n/a	n/a		1.3	12.8	85.9	n/a	n/a	n/a
1968	n/a	n/a		1.2	11.9	86.9	n/a	n/a	n/a
1969	n/a	n/a		1.1	10.2	88.7	n/a	n/a	n/a
1970	n/a	n/a		1.2	9.6	89.2	n/a	n/a	n/a
1971	0.7	8.4	90.8	1.1	9.4	89.5	2.2	11.4	86.4
1972	0.6	7.9	91.6	1.1	8.5	90.4	1.9	11.7	86.4
1973	0.3	6.2	93.5*	1.0	10.1	88.9	2.4	10.5	87.0
1974	0.4	6.4	93.2*	1.0	11.6	87.4	2.2	11.0	86.8
1975	0.5	99.5*		0.8	11.2	88.0	1.2	9.6	89.2
1976	0.3	99.7*		0.9	10.5	88.6	1.7	11.6	86.7
1977	0.4	99.6*		0.8	10.7	88.5	1.6	13.0	85.4
1978	n/a	n/a	n/a	n/a	n/a	n/a	n/a	n/a	n/a
1979	n/a	n/a	n/a	1.1	10.9	88.0	n/a	n/a	n/a
1980	n/a	n/a	n/a	n/a	n/a	n/a	n/a	n/a	n/a
1981	n/a	n/a	n/a	1.1	12.4	86.5	1.1	99.8	
1982	n/a	n/a	n/a	n/a	n/a	n/a	n/a	n/a	n/a
1983	n/a	n/a	n/a	1.1	25.0	73.9		13.4	
1984	n/a	n/a	n/a	1.5	19.1	79.3		13.7	

Legend

* Does not include other transportation equipment
BR = Basic Research; AR = Applied Research; D = Development
n/a = Not available

SOURCE: Percentages calculated on data published by National Science Foundation, *R&D in Industry*, various years.

Furthermore, this preeminence of chemicals with respect to research performance is not a recent development. This sector has been the most research-intensive sector of the American economy throughout the twentieth century. If research intensity is measured by the employment of scientific personnel (scientists and engineers) expressed as a percentage of total employment, occasional surveys conducted by the National Research Council indicate that the chemical sector's research intensity was more than twice as great as any other sector between 1921 and 1946.[1]

An understanding of the present state of this industry, in terms of how individual countries rank with respect to performance and commercial success, requires some historical perspective. America's considerable success in this industry in recent decades has to be understood against the background of international differences in natural resource endowments and the working out of what economists call path-dependent phenomena. Because the United States around the turn of the century already had an important domestic petroleum industry, and Great Britain, Germany, and France had essentially no petroleum supplies of their own, the United States readily, and at an early date, switched to a petrochemical base. The switch in resources was full of consequences, because experience with petroleum and petroleum refining led to the acquisition of many skills and capabilities that were, later, readily transferable to other chemical processing activities. This story, of the acquisition of skills and concepts that were acquired in petroleum, and their subsequent transfer to other large-scale continuous processing industries, is a central theme of the historical process by which America gained a position of world leadership. But this emergence had its base in differences in natural resource endowments and the consequences that flowed from that initial difference. This is where path-dependence became crucially important. The abundance of a particular resource at a particular point in historical time set in motion a movement, the direction and momentum of which had consequences that persisted even when the forces that gave rise to that movement had receded.

On the other hand, an important aspect of the emerging discipline of chemical engineering is that it may also offer ways of exploiting alternative, lower-cost materials in the production of new or old products. The Haber/Bosch process, the first great milestone of chemical engineering, involved a new way of producing a very old product—ammonia. But it did so by shifting the underlying German resource base from a limited resource—the by-product ovens of the iron and steel industry—to an immensely abundant base—atmospheric nitrogen.

There is an interesting counterpoint to these historical developments. On the one hand, the U.S. abundance of petroleum gave rise to a whole set of path-dependent phenomena by shifting U.S. industry to dependence on a resource, petroleum, that was available in abundance. On the other hand, the Haber/Bosch process, emerging in the second decade of the twentieth century, was a supreme instance of a country developing a new technology that enabled it to overcome the shortage of a critical industrial input—nitrogen. Thus, it is safe to say that, in these matters, history does indeed shape present capabilities very much, and many matters in which we have a current interest can be accounted for only by recourse to path-dependency types of explanation.

But neither is path-dependency the whole story, much less a simple story. What can be said is that Europe's lead in the chemicals industry, in the late nineteenth and early twentieth centuries, did not provide the most effective path for leadership in the chemical processing technology that later came to dominate the twentieth century. Whereas the United States was a distinct latecomer to the chemicals scene, its abundance of petroleum deposits and the experience that it had gained in continuous processing methods in exploiting these deposits, opened up a technology development path that provided an excellent entry into the chemical processing technologies of the mid-twentieth century. Nevertheless, that can only be a part of the story: Opening up a path in no way guarantees accelerated movement along that path. To put the matter in Toynbeesque terms, challenges sometimes generate vigorous responses; but sometimes they also overwhelm and prove to present insurmountable barriers.[2]

It is important to grasp the several separate dimensions along which productivity improvements are generated by innovations in chemical processing.

1. There are the major, Schumpeterian innovations that occur relatively infrequently but, when they do, they open up a wide range of significant new opportunities at substantially higher levels of productivity. The Haber/Bosch process is an excellent example of such a major innovation. But chemical innovations not only raise productivity in the conventional sense. They may also offer products that are not only of better quality but are more precisely configured and differentiated to cater more effectively to specific categories of consumer needs.

2. There is a flow of productivity and capacity improvements associated with the use of each of the major innovations. These improvements essentially involve a growing familiarity with a new technology once it has been introduced. Their impact is captured in the declining slope of learning curves or discussed in the vast "learning by doing" literature and popularized by the publications of the Boston Consulting Group (see Figure 1). However, a smooth movement down these learning curves may be interrupted by subsequent major innovations that offer the possibility of moving to drastically new, cost-reducing technologies.

3. There is also a continual flow of individually small design improvements and modifications within the basic framework of indi-

vidual Schumpeterian innovations (A_1, A_2, and A_3 of Figure 1). These have the effect of offering superior technologies to firms that are prepared to make the necessary investment in equipment embodying the latest designs and modifications of earlier major innovations that have experienced this subsequent improvement process. Many of these improvements are the outcome of what is essentially a "learning by using" process. That is to say, there are many ways of improving the design and operation of new equipment that become apparent only by observing difficulties or opportunities that emerge during the actual operation of the new equipment.[3]

Obviously, these small, continuous improvements in design and components become possible only after major, Schumpeterian inno-

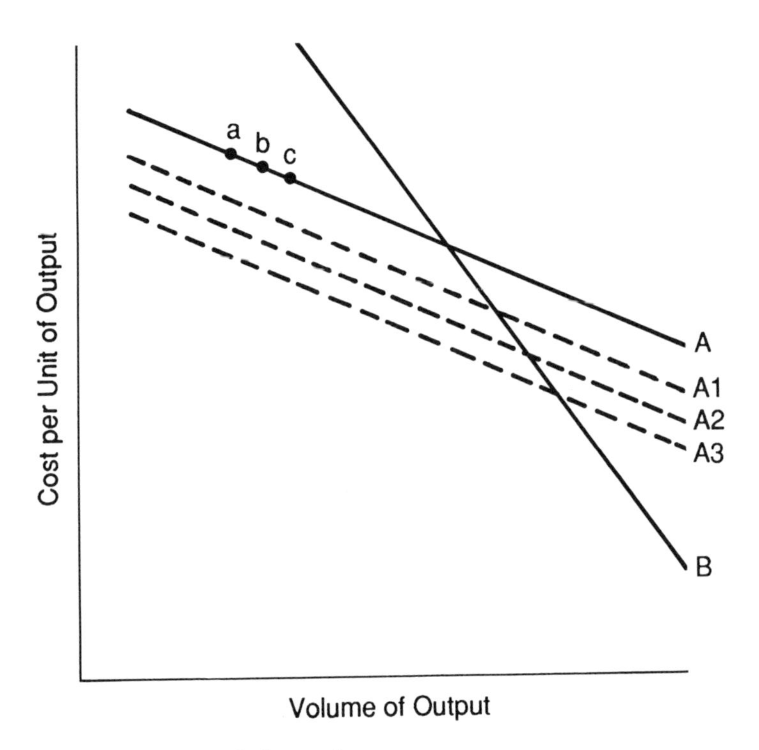

FIGURE 1 Learning curves in innovation

Legend

A = Present plant and technology
a,b,c = Movements down the learning curve of present plant
A_1,A_2,A_3 = Minor, continuous improvements embodied in new plants
B = Learning curve associated with major innovation

vations have occurred. Such improvements do not—cannot—take place in a vacuum. They are, rather, improvements on a prior innovation that provides a new framework of opportunities; they do not occur independently of such innovations. The essential point is that major innovations set the stage and provide the specific context and opportunities for the smaller, subsequent improvements process. There is much evidence that the cumulative importance of these individually small improvements is immensely important to productivity growth. Unfortunately, it is an aspect of the innovation process that has been badly neglected.

The overwhelming emphasis that has been placed, in recent years, on moving down the learning curve of an existing, unchanging plant [category (2)], fails to take account of the steady flow of incremental improvements in plant design [category (3)] that, at some point, makes it economically attractive to introduce new facilities incorporating these later improvements. Thus, a more complete depiction of the competitive process in this industry is that there is a simultaneous movement on two fronts: (a) The technological frontier, originating with a major innovation, is being continually pushed out, as design and component improvements become available and offer competitive advantages to adopters of this latest technology. This is represented in Figure 1 as learning curves shift inward toward the origin—A_1, A_2, A_3 etc.; (b) Firms have the opportunity of moving down the traditional learning curve established by their existing plant and equipment. But it should now be apparent that it is a serious mistake to visualize the competitive process as if it were entirely a matter of squeezing out, as rapidly as possible, the cost reductions offered by such existing learning curves. This is because the ongoing changes in designs and components mean that the well-known learning curve improvements take place on technologies that are, themselves, quickly becoming at least slightly obsolete. In this industry the rapid rate of technological change means that the economic life of a technology is commonly rendered obsolete long before its useful life is exhausted and, perhaps, also long before the firm has been able to approach the lower asymptote of its existing learning curve. Thus, a critical decision is to determine when it becomes worthwhile to commit to an investment that will replace the existing technology with the newest technology. There is an easy formal answer that is provided by economic analysis, which states that firms ought to continue to operate existing technology so long as it covers its marginal costs by doing so. This is, however, only a very inadequate short-run answer in the context of an industry undergoing rapid—and uncertain—technological change.

Thus, the fundamental tension in chemical processing plant is this:

Adopting the newest technology requires a huge financial commitment in physical and intangible assets of a long-lived nature. Once such an asset is acquired, learning curve improvements make it possible to raise productivity, reduce costs, and perhaps also raise product quality from this equipment. At the same time, however, the steady forward movement of the technology frontier means that it is often possible for a later entrant to start with an equipment base which begins at a cost level that may be lower than that of the earlier entrant. Nevertheless, partially for the reason already mentioned, if the earlier entrant has had the opportunity to move rapidly down his learning curve and gain a commanding market share, the new entrant may not be able to dislodge him.

But even this statement understates the inevitable uncertainties and surprises that characterize the innovation process in chemical processing. On the one hand, as already suggested, technical improvements commonly occur before an innovative process has moved very far down its potential learning curve. This is particularly poignant since many promising new technologies are promising precisely because they offer the prospect of sharply declining learning cost curves, but they must nevertheless begin their productive lives at cost levels that may be even higher than the present costs of technologies already in existence (In Figure 1, the upper portion of learning curve B). Finally, sudden shocks, such as a sharp rise in energy costs, or the availability of a cheaper feedstock, can lead to a rapid redefinition of what constitutes an optimal technology. In an industry of long-lived and expensive assets, these uncertainties render the investment decision an especially painful process—one need only recall the years immediately following the oil boycott by the Arab members of OPEC in 1973.

As the chemical industry has grown and matured, it has given rise to an entirely new specialization: The discipline of chemical engineering, which simply did not exist a hundred years ago. The chemical engineer has become the critical factor in taking the products of the research process and developing feasible techniques for producing them on a commercial basis. It must be emphasized that the findings of laboratory research do not provide the information necessary for commercial production. Such production is not a matter of simply scaling up the tubes and retorts in which a new product was originally developed. That is often physically impossible and hardly ever economically sensible.

Nor is chemical engineering reducible to applied chemistry. It could be better described as the application of mechanical engineering to production activities involving chemical processing. The essence of chemical engineering, then, is a cluster of integrative skills that are applied to the design of chemical processing equipment. But there is much more to it than that. The chemical engineer has, at the center of his activities, the examination and synthesis of different technologies from the point of view of their comparative cost. The work and decision making of the chemical engineer is inherently economic as much as it is engineering, since it involves the explicit consideration of innumerable tradeoffs in determining optimal design.

Moreover, it is clear from what has already been said that success in the commercialization of chemical processing innovations has depended critically upon the productivity gains realized through an improvement process that takes place after an innovation is introduced in the market. An integral part of this process of cumulative improvement, which deserves separate recognition and treatment, has involved the exploitation of economies of large scale production and therefore a movement toward larger scale plants.

Historically, success in the commercialization of new technologies in this sector has turned upon the ability to make the transition from small scale, batch production to large scale, continuous processing plants. The benefits of larger scale have been so pervasive in this sector that chemical engineers have developed and employed a "six-tenths rule," which is regularly invoked, that is, capital costs increase by only 60 percent of the increase in rated capacity.

A distinctive characteristic of the American chemical processing scene even in its earliest years was the continuous pressure toward the exploitation of larger size, and the alacrity with which American firms moved in that direction. One authoritative study, discussing the American situation shortly before its entry into the First World War, has referred to " . . . the American attitude to the size of chemical works, which was, in short, to build a large plant and then find a market for the products."[4] It would seem plausible to infer that such an attitude developed at the time because the relevant markets were, as a matter of fact, both large and growing rapidly.

As the industry shifted to petroleum feedstocks in the interwar years and mastered the problems of large-scale, continuous process operations, the optimal size of plant often grew to exceed the market requirements of even the largest of western European countries. Since the European industry had relied much more heavily in its earlier years upon coal as the basic raw material, the transition to larger scale was impeded by skills, attitudes and educational preparation that had been developed under that coal-based industrial regime. European developments were also influenced by the determination of each country to maintain a capability for satisfying the requirements of its own domestic market. "Even in countries with a relatively large population, such as France and Great Britain, chemical firms planning new projects in the postwar period found it difficult to build a large enough plant that would have reasonably attractive economics. Substantial exports were needed to build such plants, but the products in question would not necessarily be saleable in adjacent European countries, since potential purchasers were still averse to being dependent on supply from across the border."[5]

Building larger chemical processing plants is, however, much more than merely having assurance of access to sufficiently large markets. Such larger plants are necessarily also a product of technological innovations that make them feasible. In this respect it is much more common than it ought to be to assume that the exploitation of the benefits of large scale production is a separate phenomenon independent of technological change. In fact, larger plants typically incorporate a number of technological improvements, based upon the wealth of experience and insight into better plant design, that could be accumulated only through prolonged exposure to the problems involved in the operation of somewhat smaller plants. The building of larger plants must, as a result, often await advances in the technological capabilities in plant design, equipment manufacture, and process operation. Thus, the benefits of scale cannot be attained until certain facilitating technological conditions have been fulfilled.

Both as a conceptual matter and as a practical matter, it is not easy to disentangle the benefits of larger scale production from those achieved through introduction of improved equipment, improved design, or better "know-how," that is, better understanding of the technological relationships that are eventually embodied in the larger plant. Such later plants typically incorporate a large number of cumulative improvements and conceptual insights.

The discussion of scale raises a final set of considerations. Scale factors have been important not just at the level of the individual plant and its optimal output, compared to the size of the available market. A central additional question is whether the market is large enough to support specialist plant contractors and designers who will eventually be responsible for delivering the plant and the equipment. This is a critical and badly neglected consideration, because the chemical sector has developed a unique set of specialist firms and organizations which have, in turn, played a major role in the innovation process. These specialists now operate on a world scale for a world market, and commercial success and failure must inevitably be addressed in terms of that world market and the ability of various specialist firms to prevail against competition in that market.

Specialized engineering firms (SEFs) came to play a critical role in the chemicals sector during the years following the Second World War. Chemical firms had subcontracted functions like procurement and installation to the SEFs even before the War, when design and process development was essentially carried out in-house. Chemical companies typically carried out their own process design, and used external contractors to handle construction, piping and mechanical work, electrical work, and other separate facets of the project. Petroleum companies typically farmed out most of the detailed design as well to SEFs. After the war, the chemical firms increasingly relied upon SEFs to design, engineer, and develop their manufacturing installations. In the 1960s, nearly three-quarters of the major new plants were engineered, procured, and constructed by specialist plant contractors."[6]

There were various specific advantages accruing to SEFs in designing and developing chemical production processes. First, during the 1920s and 1930s, while large chemical companies had concentrated mainly upon product innovation and development, SEFs had acquired an ability to handle sophisticated process design and development work. In this, they had benefited greatly from their experience in the petroleum sector, which had faced, earlier than the chemical industry, problems of large scale processing and refining. The unique capabilities derived from this earlier involvement in design and development work for the petroleum sector constitutes a critical instance of the role of path-dependent phenomena, referred to earlier. As the world moved into the petrochemical age, some countries were better situated by their own past for dealing with the new design and production problems of the new chemical industry. History indeed matters.

A further important source of advantage to SEFs came from their opportunity for exploiting economies of specialization and certain forms of learning by doing. Once a major new process technology was developed, or the scaling up of a given production process was carried out, SEFs could reproduce that new technology, or larger scale production process, for many clients. Such economies could not be accumulated by the chemical manufacturers themselves, precisely because they could produce that technology only for their own, limited internal needs, whereas SEFs had a much more extensive experience with designing that particular plant many times for different clients. Moreover, as they worked for many different clients, they accumulated useful information related to the operation of plants under a variety of conditions. This represented an opportunity for accumulating knowledge and specialized skills which were not available to the chemical producers. SEFs thus acquired the capability to design better plants for other potential customers.

The role of SEFs had important consequences with respect to competition among chemical manufacturers on a global scale. The most significant was their development of the complete technology and plant designs for the basic building blocks of the chemical industry, for example, olefins and aromatics. American-designed ethylene cracking plants appeared all over the world, and these in turn required technologies for the manufacture of key intermediates for the chemical industries of many nations. Such technologies were supplied by SEFs and manufacturers in other countries who sought to generate additional revenues outside their own domestic markets. Latecomers to a particular chemical technology could benefit from their relations with SEFs, which were able to provide them with the process know-how that they had accumulated, at least in part, through their previous relations with earlier entrants (Thus, in terms of Figure 1, latecomers were likely to be supplied with plant that incorporated the design improvements designated by broken lines A_1, A_2, and A_3). Moreover, the availability of such technology from SEFs also encouraged many new entrants into the industry from related sectors such as petroleum, paper, food, metals, and the like. A result was intensified competition, including periods of overbuilding and excess capacity.

In the postwar period, then, the world chemical industry was powerfully shaped by successive waves of diffusion of new technologies, including both product and process technologies. Although the sources of chemical innovation were diverse, a major factor was the role played by American specialized engineering contractors. More specifically, the division of labor between SEFs and the chemical manufacturers had important consequences for the diffusion of new technology, both at domestic and international levels. SEFs licensed extensively to chemical firms all over the world. As a result, they served as major carriers of technological capabilities, including highly elusive but significant "know-how," that is, essential knowledge of a noncodified sort that was, nevertheless, vital to successful plant operation and performance.

The vital role played by SEFs in designing and diffusing new technologies in the chemicals sector underlines a point that it is useful to make in closing. That is, the competitive process, even in high tech industries, needs to be examined in terms of a range of activities located "downstream" from the scientific research process. Economists have not, so far, done a very thorough job of this. They have, on the whole, treated technological innovation in a highly abstract way as a collection of activities going on inside a black box, the contents of which are never subjected to systematic examination. When the inputs into that black box are unpacked, it turns out that R&D expenditures are, in fact, not primarily spent on scientific research, but on development which, in the United States has, for many years, constituted more than two-thirds of all R&D spending. Alternatively, even where scientists dominate the initial stages of new product development, the later stages, and eventual commercial success, are likely to be dominated by engineering, design, and technological capabilities.

If these "downstream" activities seem to be lacking in glamour and to be, in fact, rather pedestrian, no doubt they are, at least from certain perspectives. But that perspective is likely to belong to the academic or the intellectual, who is interested in "the big picture" or in large conceptual breakthroughs. It is essential to understand that the marketplace renders judgments that are based on modest improvements and the cumulative effect of individually small, pedantic modifications in product or process design. Small, incremental improvements have brought the semiconductor industry from a handful of transistors on a chip to more than a million such transistors; in telecommunications, it has brought the channel capacity of a 3/8-inch coaxial cable to more than an order of magnitude increase over an earlier level; and in the computer industry the speed of computational capability has been increased, by individually small increments, by many orders of magnitude. In high-tech as well as in low-tech industries, an unkind Providence seems to have ordained that commercial success is likely to favor particularly the possessors of a varied assortment of grubby skills.

NOTES

1. See David Mowery and Nathan Rosenberg, *Technology and the Pursuit of Economic Growth*, Cambridge University Press, New York, 1989, pp. 64-71.

2. See Peter Spitz, *Petrochemicals*, John Wiley and Sons, New York, 1988, pp. xiii, 26-29, and 57-60.

3. These design and component improvements can sometimes be installed or retrofitted into existing equipment, but usually at a higher cost than when they are introduced at the stage of the actual manufacture of new equipment. In other cases new components can sometimes be installed during normal maintenance and replacement activities. See Ralph Landau (ed.), *The Chemical Plant*, Reinhold Publishing Corporation, New York, 1966. For further discussion of learning by using, see Nathan

Rosenberg, *Inside the Black Box*, Cambridge University Press, New York, 1982, chapter 6.

4. L. F. Haber, *The Chemical Industry 1900-1930*, Oxford University Press, Oxford, 1971, p. 176.

5. Spitz, *Petrochemicals*, op. cit. p. 348. See also Ralph Landau, "Chemical Engineering in West Germany," *Chemical Engineering Progress*, July 1958.

6. C. Freeman, "Chemical Process Plant: Innovation and the World Market," *National Institute Economic Review*, #45, August 1968.

Ralph Landau owns Listowel, Inc., and is a consulting professor of economics at Stanford University and a faculty fellow at the Kennedy School of Government at Harvard University. He codirects programs in technology and economic growth and policy at both institutions. The holder of an Sc.D. degree in chemical engineering from M.I.T., Dr. Landau in 1946 cofounded Halcon International, a chemical engineering firm that he headed for 36 years; 20 years later he cofounded the Oxirane Group with ARCO. He is a past vice president of the National Academy of Engineering, and in 1985 was among the first recipients of the National Medal of Technology. His other awards include the Perkin and Chemical Industry Medals, AIChE's Founders Award, and the John Fritz Medal.

Nathan Rosenberg is a Fairleigh S. Dickinson, Jr. Professor of Public Policy in the Department of Economics, Stanford University. He is past chairman of the Department of Economics and is the director of the Technology and Economic Growth Program in Stanford's Center for Economic Policy Research. He is the author of numerous articles and several books focusing primarily on the economics of technological change. He has also served on the faculty at the University of Wisconsin, Harvard University, Purdue University, and the University of Pennsylvania. Dr. Rosenberg earned his B.A. degree from Rutgers University and his M.A. and Ph.D. degrees from the University of Wisconsin.

List of Symposium Participants

John A. Alic
Office of Technology Assessment
U.S. Congress

Tom Althuis
Director
Science Policy
Pfizer, Inc.

Dieter H. Ambros
Executive Board
Henkel KG AA

Jesse Ausubel
Carnegie Commission on Science, Technology, and Government
Rockefeller University

Martin Neil Baily
Professor
University of Maryland

Jordan Baruch
President
Jordan Baruch Associates

Marlene R.B. Beaudin
Associate Executive Director
Commission on Engineering and Technical Systems
National Research Council

William B. Beeman
Vice President
Macro Economics
Committee for Economic Development

Charles M. Benbrook
Executive Director
Board on Agriculture
National Research Council

Jules Blake
Vice President for
Corporate and Science Affairs
Colgate Palmolive Company

Erich Bloch
Director
National Science Foundation

Mark A. Bloomfield, Esq.
President
American Council for Capital Formation

Elkan R. Blout
Dean for Academic Affairs
Harvard School of Public Health

David Bodde
Executive Director
Commission on Engineering and Technical Systems
National Research Council

Joseph Bordogna
Dean of Engineering
University of Pennsylvania

Michael J. Boskin
Chairman
Council of Economic Advisers

Michel Boudart
Department of Chemical Engineering
Stanford University

John Brigden
Financial Analyst
General Electric Company

Alfred E. Brown
Retired Director
Scientific Affairs
Celanese Corporation

David Brown
Retired Vice Chairman
Halcon International, Inc.

Robert A. Brown
Department Head
Department of Chemical Engineering
Massachusetts Institute of Technology

Solomon J. Buchsbaum
Executive Vice President
Customer Systems
AT&T Bell Laboratories

L. Gary Byrd
Consulting Engineer

Louis W. Cabot
Chairman of the Board
Brookings Institution

Gordon Cain
Chairman
The Sterling Group

Jean H. Carter
The Business Council

Joseph V. Charyk
Retired Chairman and CEO
COMSAT

David Cheney
Council on Competitiveness

Hirsh Cohen
Alfred P. Sloan Foundation

Justin W. Collat

Paul M. Cook
Chairman and Chief Executive
Raychem Corporation

William D. Craig

Michael R. Darby
Assistant Secretary for Economic Policy
Department of the Treasury

Paul David
Department of Economics
Stanford University

Charles H. Davidson
Foundation Director
American Society for Industrial Security

Gerald P. Dinneen
Foreign Secretary
National Academy of Engineering

Thomas E. Everhart
President
California Institute of Technology

Alan E. Fechter
Executive Director, Office of Scientific and Engineering Personnel
National Research Council

Daniel J. Fink
President
D. J. Fink Associates, Inc.

Kenneth Fulton
Special Assistant to the Chairman
National Research Council

Melvin J. Gipson
Program Associate
Program Office
National Academy of Engineering

Ralph E. Gomory
President
The Alfred P. Sloan Foundation

Mary L. Good
President
Engineered Materials Research
Allied-Signal Inc.

William E. Gordon
Foreign Secretary
National Academy of Sciences

William Gorham
President
Urban Institute

David Goslin
Director
American Institute for Research

Bruce Guile
Director
Program Office
National Academy of Engineering

Milton Harris
Retired Vice President for R&D
Gillette Company

George N. Hatsopoulos
Chairman and President
Thermo Electron Corporation

Michael Heylin
Editor
Chemical and Engineering News

Joseph S. Hezir
Deputy Associate Director of Energy and Science Division
Office of Management and Budget

Christopher T. Hill
Director
Manufacturing Forum
National Academy of Engineering
National Academy of Sciences

Allan R. Hoffman
Director
Academy Industry Program
National Research Council

John D. Holmfeld
Committee on Science, Space, and Technology
U.S. House of Representatives

William G. Howard, Jr.
NAE Senior Fellow
National Academy of Engineering

Barbara Huff
Executive Associate
NAE Executive Office
National Academy of Engineering

Kathryn Jackson
NAE Fellow
National Academy of Engineering

Jeanne Jacob
Director
Development Office
National Academy of Engineering

Walter Joelson
Chief Economist
General Electric Company

Trevor O. Jones
Chairman
Libbey-Owens-Ford, Inc.

Dale W. Jorgenson
Frederick Eaton Abbe Professor
Department of Economics
Harvard University

Maribeth Keitz
Program Assistant
Program Office
National Academy of Engineering

Ronald E. Kutscher
Associate Commissioner
Office of Economic Growth and Employment Projections
Bureau of Labor Statistics

Ralph Landau
Consulting Professor of Economics
Stanford University

Laurie Landeau

H. Dale Langford
Editor
National Academy of Engineering

Charles F. Larson
Executive Director
Industrial Research Institute

William E. Leonhard
Chairman, President and CEO
The Parsons Corporation

Carrie Levandoski
Director
Administration, Finance, and Public Awareness
National Academy of Engineering

Robert E. Litan
Director
Brookings Institution

John W. Lyons
Director
National Institute of Standards and Technology

Bruce K. MacLaury
President
Brookings Institution

Plato Malozemoff
Chairman Emeritus
Newmont Mining Corporation

Robert Malpas
Managing Director
Powergen

Virginia Martin
Publisher
Scientific and Technical Division
John Wiley and Sons, Inc.

Edward A. Mason
Retired Vice President - Research
Amoco Corporation

Lawrence McCray
Executive Director
NRC/COSEPUP

John L. McLucas
Retired Chairman
QuesTech, Inc.

Charles W. McMillion
Institute for Policy Studies
Johns Hopkins University

Stephen Merrill
Associate Executive Director
Office of Government and External Affairs
National Research Council

Joseph Minarik
Executive Director
Joint Committee on Economics
U.S. Congress

W. David Montgomery
Assistant Director
Natural Resources
Congressional Budget Office

J. David Morrissy
Office of U.S. Trade Representative

Ed Muller

Richard R. Nelson
Columbia University

Richard S. Nicholson
Executive Officer
American Association for the Advancement of Science

Courtland D. Perkins
Past President
Consultant
National Academy of Engineering

Don I. Phillips
Executive Director
GUIR Roundtable
National Academy of Sciences
National Academy of Engineering
Institute of Medicine

Jonathan Piel
President and Editor
Scientific American

Gail Porter
Director
Office of News and Public Information
National Research Council

Frank Press
President
National Academy of Sciences

Victor Rabinowitch
Executive Director
Office of International Affairs
National Research Council

Jon Rehnberg

Dorothy Robyn
Joint Economic Committee
U.S. Congress

David A. H. Roethel
Executive Director
American Institute of Chemists, Inc.

Barry Rogstad
President
American Business Conference

Nathan Rosenberg
The Faculty of Economics
Cambridge University

Richard Rosenbloom
Professor
Harvard Business School

Michael B. Rubin
House Subcommittee, Transportation, Aviation, and Materials
U.S. House of Representatives

Allen S. Russell
Retired Vice President and Chief Scientist, ALCOA
Retired Adjunct Professor
University of Pittsburgh

Vernon W. Ruttan
Department of Agricultural and Applied Economics
University of Minnesota, St. Paul

Alfred Saffer
Retired Vice Chairman
The Halcon S.D. Group, Inc.

William C. Salmon
Executive Officer
NAE Executive Office
National Academy of Engineering

Roland W. Schmitt
President
Rensselaer Polytechnic Institute

Claudine Schneider
Congressional Competitiveness Caucus
U.S. House of Representatives

Elliot Schwartz
Congressional Budget Office

Robert C. Seamans, Jr.
Senior Lecturer
Department of Aeronautics and Astronautics
Massachusetts Institute of Technology

Martin Siegel
Director
Government Relations
American Institute of Chemical Engineers

Hedy E. Sladovich
Program Associate
Program Office
National Academy of Engineering

William J. Spencer
Vice President
Corporate Research Group
Xerox Corporation

H. Guyford Stever
Corporate Director and Science Advisor

Lawrence Summers
Professor of Economics
Harvard University

Morris Tanenbaum
Vice Chairman
AT&T Company

Gerald F. Tape
Retired President
Associated Universities, Inc.

Annmarie Terraciano
Program Assistant
Program Office
National Academy of Engineering

Samuel O. Thier
President
Institute of Medicine

Charls E. Walker
Chairman
Charls E. Walker Associates

David Walters
Chief Economist
Office of the U.S. Trade Representative

Zhenquan Wang
First Secretary for Science and Technology
The Chinese Embassy

John F. Welch, Jr.
Chairman and CEO
General Electric Company

Albert R. C. Westwood
Vice President
R&D Development
Martin Marietta Corporation

Robert M. White
President
National Academy of Engineering

F. Karl Willenbrock
Assistant Director, Scientific, Technological and International Affairs
National Science Foundation

David Williams
Comptroller
National Research Council

Pat Windham
Professional Staff Member
Committee on Commerce, Science, and Transportation
U.S. Senate

Julia Wolf
Development Associate
Development Office
National Academy of Engineering

Douglas Wolford
Public Awareness Associate
National Academy of Engineering